Moritz Scharff

Bio-inspired Tactile Sensing

Bio-inspired Tactile Sensing

Analysis of the inherent characteristics of a vibrissa-like tactile sensor

Moritz Scharff

Universitätsverlag Ilmenau
2021

Impressum

Bibliografische Information der Deutschen Nationalbibliothek
Die Deutsche Nationalbibliothek verzeichnet diese Publikation in der Deutschen Nationalbibliografie; detaillierte bibliografische Angaben sind im Internet über http://dnb.d-nb.de abrufbar.

Diese Arbeit ist in einem gemeinsamen Promotionsverfahren mit der Pontificia Universidad Católica del Perú und der Technischen Universität Ilmenau entstanden und hat der Fakultät für Maschinenbau der Technischen Universität Ilmenau als Dissertation vorgelegen.

Tag der Einreichung:	14. Juli 2020
1. Gutachter:	Prof. Dr.-Ing. habil. Carsten Behn (Technische Universität Ilmenau, Hochschule Schmalkalden)
2. Gutachter:	Prof. Dr. Jorge Hernan Alencastre Miranda (Pontificia Universidad Católica del Perú)
3. Gutachter:	Univ.-Prof. Dipl.-Ing. Dr. med. habil. Hartmut Witte (Technische Universität Ilmenau)
Tag der Verteidigung:	18. Dezember 2020

Technische Universität Ilmenau/Universitätsbibliothek
Universitätsverlag Ilmenau
Postfach 10 05 65
98684 Ilmenau
http://www.tu-ilmenau.de/universitaetsverlag

ISBN 978-3-86360-237-6 (Druckausgabe)
DOI 10.22032/dbt.47879
URN urn:nbn:de:gbv:ilm1-2020000405

Titelgrafik: Veit Henkel | Fakultät für Maschinenbau, TU Ilmenau

„Better bent than broken.“
Steven Vogel in Drag and Flexibility in Sessile Organisms

Abstract

Tactile sensing is a fast-developing research area that is important, e.g., for autonomous robot systems in the context of path planning and navigation in uncertain terrains. One possibility for novel enhanced designs of tactile sensors is to analyze and adapt natural paragons. Rodents like rats have tactile sensory hairs, so-called vibrissae, on both sides of their muzzle. These hairs exhibit a sophisticated structure and geometrical shape. The slender, tapered, and inherent curved hair shaft is supported by a hair follicle that includes mechanoreceptors. Touching an object with their vibrissae, rats can recognize the shape of an object or determine properties of its surface texture by evaluating only the signals inside the hair follicle. The present work contributes to the overall goal to unfold the ability of natural vibrissa with view to applications in engineering like surface metrology or autonomous robots. The vibrissal system is described in detail, analyzed, and interpreted using the idea of a biomechatronic system and stimulus leading apparatus. The properties of a natural vibrissa contribute in a complex manner and in various ways to its functionality and determine its inherent characteristics. Therefore, properties of a natural vibrissa are systematically adapted to an artificial tactile sensor. Using this artificial vibrissa-like sensor, it is shown that three different kinds of information about the scanned object are present in the captured data: the overall object shape, a macroscopic, and a microscopic surface texture. The regarding information is encrypted in the signals at the base of the sensor and must be processed to recognize each type of information. To support this process, it is found that a larger distance to an object, around 80% of the length of the sensor shaft, permits a good detectability of properties regarding the macroscopic surface texture. For a moderate distance, e.g., 60% of the length of the sensor shaft, the detection of a microscopic surface texture works best. Finally, a closer distance, below 45% of the length of the sensor shaft, is advantageous for a recognition of the overall shape. These characteristics -strongly depending on the distance- are closely related to the elasticity of the sensor shaft. For example, scanning a close object with distinct macroscopic surface texture, proper detection of the macroscopic surface texture is not possible because the bending induced curvature of the sensor shaft prevents contact between the sensor shaft and small shapes of macroscopic surface features like grooves and gaps. Therefore, the deformed sensor shaft fulfills the task of a morphological filter. The diameter of the sensor defines the limit between macroscopic and microscopic surface texture. Surface texture elements smaller than this limit belonging to the microscopic surface texture. Therefore, also the tip diameter is an inherent characteristic of the sensor. A further inherent characteristic is found while scanning a microscopic surface texture and retracting the sensor shape in the way that the concave side of the deformed elastica points in the displacement direction.

This configuration and displacement cause an amplification of the recorded support reactions. Following these kinds of ideas, the inherent characteristics of the sensor system are analyzed in the present work.

Resumen

El desarrollo de sensores táctiles ha cobrado creciente importancia por su posible aplicación en robots autónomos móviles en el contexto de la exploración de ambientes desconocidos. Una posibilidad para desarrollar sensores táctiles es analizar y adaptar las propiedades esenciales de ejemplos que encontramos en la naturaleza (biomimética). Un ejemplo de esto son las vibrisas o pelos táctiles de la rata, ubicados a ambos lados del hocico del animal. Las vibrisas tienen forma de pelo largo, delgado y cónico y con una curvatura inherente. Cada vibrisa se encuentra fijada a un folículo que incluye mecanorreceptores que registran los estímulos provenientes del exterior. Durante la exploración de un ambiente, o de un objeto desconocido, la rata mueve sus vibrisas para detectar la forma del objeto o su textura superficial. Ese procedimiento induce la activación de los mecanorreceptores de los folículos, generando las señales que llevan la información sobre el objeto y permiten al animal conocer su medio ambiente. Esta tesis es una contribución para adaptar las propiedades y funciones de una vibrisa natural a un sistema tecnológico bioinspirado. Se ha realizado un análisis metodológico debido a que las propiedades y las funciones de una vibrisa tienen una relación sinérgica. En ese sentido, la vibrisa natural se ha descrito utilizando las ideas del Biomechatronic System y del Stimulus Leading Apparatus. Partiendo de ese punto, se ha diseñado un sensor artificial inspirado en una vibrisa natural con el objeto de investigar las propiedades y efectos inherentes de ese tipo de sensor táctil. Se han escaneado diferentes superficies para investigar el funcionamiento del sensor en esas condiciones. Las señales registradas mostraron que un objeto está caracterizado por su forma general y sus texturas superficiales en escalas macro- y microscópica. Esta información sobre el objeto se encuentra superpuesta en las señales y, por lo tanto, es necesario extraerlas para su evaluación. En ese procedimiento, la distancia entre el soporte del sensor y el objeto tiene una influencia importante. Los elementos de una textura superficial macroscópica se registran mejor si el objeto está colocado a una distancia de 80% de la longitud del captador del sensor. Una distancia media, 60% de la longitud del captador, es adecuada para escanear la textura superficial microscópica. Finalmente, con una distancia pequeña de menos de 45% de la longitud del captador es posible detectar la forma general del objeto. Estos efectos deben ser interpretados en relación a la elasticidad del captador. Por ejemplo, una distancia pequeña evita que elementos de la textura superficial macroscópica sean registrados porque la deformación y la curvatura correspondiente evitan que el captador pueda tocar elementos pequeños como ranuras y etapas. Por lo tanto, la deformación del captador está en función de un filtro morfológico y, en consecuencia, esa es una de las características inherentes del sensor.

La transición de la textura superficial macroscópica a la textura superficial microscópica está definida por el diámetro de la punta del captador; los elementos menores de la textura superficial corresponden a la escala microscópica. Por lo tanto, el diámetro de la punta es también una característica inherente del sensor. Se ha demostrado, además, que cuando se hace tocar al captador una muestra con un desplazamiento en dirección contraria a la curvatura del captador, las señales registradas se amplifican. Esta es, en consecuencia, otra propiedad inherente del sensor. Finalmente, en esta tesis, se ha analizado el comportamiento de un sensor artificial - inspirado en una vibrisa natural - considerando todas las condiciones anteriormente expuestas.

Kurzzusammenfassung

Die Weiterentwicklung taktiler Sensoren gewinnt an Bedeutung bspw. durch eine verstärkte Anwendung taktiler Sensoren zur Navigation in unbekannten Umgebungen von autonomen mobilen Robotern. Eine Möglichkeit taktile Sensoren weiter zu entwickeln ist, sich -wie auch schon andere Entwicklungen zeigen- der Natur zu bedienen, Vorbilder zu identifizieren, diese fundamental zu analysieren und als wesentlich befundene Eigenschaften und Funktionstüchtigkeiten zu adaptieren. Ratten besitzen auffällige Tasthaare an beiden Seiten der Schnauze, sogenannte Vibrissen. Diese sind gekennzeichnet durch einen langen, schlanken und natürlich vorgekrümmten Haarschaft mit konischem Querschnittsverlauf. Der Haarschaft wird von einem Haarfollikel gehalten, der sich unter der Haut befindet und in dem überdies Mechanorezeptoren zur Reizdetektion zu finden sind. Während der Erkundung von unbekannten Umgebungen und Objekten setzen Ratten ihre Vibrissen ein, um bspw. die Form oder Textur eines Objektes zu bestimmen, indem die Vibrisse daran entlang bewegt wird. Die Informationsaufnahme wird im Haarfollikel durch die Mechanorezeptoren realisiert. Die vorliegende Arbeit leistet einen Beitrag zum übergeordneten Ziel, die Eigenschaften und Einsatzmöglichkeiten dieses komplexen und hochentwickelten Sensorsystems der Natur für technische Anwendungen nutzbar zu machen. Die Eigenschaften einer Vibrisse sind synergetisch und beeinflussen ihre Funktionen in bedeutendem Maß. Deshalb wird das natürliche Vorbild detailliert analysiert und mit den Konzepten des biomechatronischen Systems und des Reizleitungsapparats beschrieben und auf dieser Basis ein Vibrissen-ähnlicher Sensor entworfen, um die inhärenten Eigenschaften eines solchen Sensorsystems zu untersuchen. Um die Funktionstüchtigkeit des entworfenen Protoyps, aufgebaut auf Basis der detaillierten Vorabanalyse, zu untersuchen, werden verschiedene Testobjekte, einschließlich verschiedener Oberflächenbeschaffenheiten, mit dem Sensor vermessen. Anhand der aufgezeichneten Messsignale zeigt sich, dass ein Objekt durch seine generelle Form sowie seine makroskopische und mikroskopische Oberflächenstruktur beschrieben werden kann. Die genannten Informationen überlagern sich in den gemessenen Signalen und müssen für eine weiterführende Auswertung extrahiert werden. Der Abstand zwischen Sensorlagerung und Objekt hat entscheidenden Einfluss. Makroskopische Oberflächenelemente lassen sich im Abstand von 80% der Länge des Sensorschafts besonders gut detektieren. Ein mittlerer Abstand, ca. 60% der Länge des Sensorschafts, unterstützt die Erfassung der Eigenschaften einer mikroskopischen Textur. Hingegen ist ein kleiner Abstand von 45% der Sensorschaftlänge besonders geeignet zur Detektion der generellen Form des Objekts. Diese Effekte sind in enger Verbindung zur Elastizität des Sensorschafts zu interpretieren.

Beispielsweise verhindert die starke Krümmung des Sensorschafts in Folge eines kleinen Objektabstandes eine Detektion von makroskopischen Oberflächenelementen wie Rillen und Stufen, da sich der Sensorschaft in dieser Konfiguration wie ein adaptiver morphologischer Filter auswirkt und damit eine inhärente Eigenschaft des Sensorsystems ist. Der Übergang zwischen der makroskopischen und mikroskopischen Oberflächenstruktur wird durch den Durchmesser der Spitze des Sensorschafts bestimmt. Oberflächenstrukturelemente, die kleiner als dieser Durchmesser sind, gehören zur mikroskopischen Oberflächenstruktur. Daraus folgt, dass auch der Durchmesser der Spitze eine inhärente Eigenschaft ist. Eine weitere inhärente Eigenschaft wird bei der Detektion einer mikroskopischen Oberflächenstruktur erkennbar. Wenn der, sich im Kontakt befindende, stark verformte, Sensorschaft so bewegt wird, dass die konkave Seite des Sensorschafts in Bewegungsrichtung zeigt, werden die erfassten Signale verstärkt - im Vergleich zur entgegengesetzten Bewegungsrichtung. Unter Berücksichtigung der genannten und weiterer Ideen, wird der durch eine natürliche Vibirsse inspirierte Sensor in der vorliegenden Arbeit untersucht.

Acknowledgement

I like to gives thanks to all who contributed in some or another way to the present work.

Special thanks go to my supervisors Prof. Carsten Behn and Prof. Jorge Hernan Alencastre Miranda. Both were exposed to a torrent of ideas and discussions and never gave up hope, in contrast, they gave me a telling-off and kept me focused.

I am grateful for the support of Prof. Klaus Zimmermann and my colleagues of the Technical Mechanics Group, Department of Mechanical Engineering, Technische Universität Ilmenau. His collaborative project „Technische, nicht-visuelle Charakterisierung von Substratkontakten nach dem biologischen Vorbild carpaler Vibrissen“ (together with Prof. Hartmut Witte und PD Dr. Manuela Schmidt, founded by Deutsche Forschungsgemeinschaft DFG: Zi 540/16-1 .../16-2, WI 1664/4-1... 4-2, SCHM 1748/7-1... 7-2) and its outcome was the best possible starting point and formed ideal experimental conditions to process this thesis at hand.
This work would be not the same without the inspiring and motivational discussions with Prof. Joachim Steigenberger, Prof. Hartmut Witte, Dr. Philipp Schorr, Thomas Helbig, Danja Voges†, Dr. Christoph Will, Maximilian Darnieder and Dr. Ana L. Albarracín.
Also, I like to acknowledge the support of the teams of the Laboratorio del Centro de Caracterizaciń de Materiales (CAM PUCP) and of the Sala de Manufactura Digital VEO 3D of the Pontifical Catholic University of Peru.
Furthermore, I give thanks to the Thüringer Graduiertenförderung for the two-year funding period of their grant to support doctoral studies.
Equally, I am indebted to the German Academic Exchange Service (DAAD) for the funding regarding several stays in Lima/Peru at the Pontifical Catholic University of Peru. Moreover, I appreciate the half-year grant of the Technische Universität Ilmenau to finish this work.

Acronyms & Symbols

FSC	follicle-sinus complex
MACF	macroscopic surface feature
MICT	microscopic surface texture

d	diameter of the sensor shaft
E	Young's modulus
$\vec{F}$	contact force
f	frequency
$\vec{F}_n$	normal component of the contact force
$\vec{F}_t$	tangential component of the contact force
$\vec{F}_x$	support force in x- direction
$\vec{F}_y$	support force in y- direction
G	shear modulus
I_z	second moment of area with respect the z-axis
L	length of the sensor shaft
$\vec{M}_{bz}$	bending moment with respect to the z-axis (stress resultant)
$\vec{M}_z$	support bending moment with respect to the z-axis
P_x	x-coordinate of the reconstructed point
P_y	y-coordinate of the reconstructed point
Re	yield strength
s	arc length of the beam
t	time
$\vec{v}$	velocity of the sensor support
v	displacement with respect to the y-axis
x_0	x-coordinate of the clamping
x_c	x-coordinate of the contact point
y_0	y-coordinate of the clamping
y_c	y-coordinate of the contact point
α	slope of the object shape
β	stiffness proportional material damping factor
η	distance between the sensor support and a reference point on the object shape
κ	curvature of the beam
κ_0	inherent curvature of the beam
λ	eigenvalue
μ	friction coefficient
μ_d	dynamic friction coefficient
μ_s	static friction coefficient
ν	Poisson's ratio
ω	angular natural frequency

ρ	density
ϑ	taper ratio
φ	slope of the beam
φ_0	angular position of the support
φ_c	slope of the beam at the contact point
ζ	friction angle

Contents

1 Introduction

The design as well as the optimization of, e.g., technologies, products, and processes influence human life every day and in every aspect. Engineers develop and implement new approaches and ideas in order to fulfill new and advanced requirements of the technical products. For example, in the scope of Industry 4.0, production processes should be fully automated and autonomous [BMWi; 2018]. In doing so, information about, e.g., the present state of the product and the environmental conditions must be captured, processed, and evaluated by artificial systems. These systems are increasingly inspired by nature [Lepora et al.; 2013]. In current literature, a method to adapt a natural system into an artificial one, or to interpret a natural paragon out of the perspective of engineering, is known as bionics, biomimetics, biomimicry, biomechatronics, or bio-inspired engineering among others.

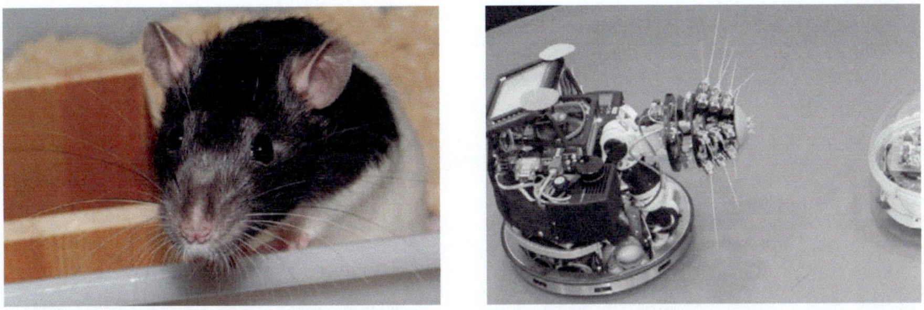

(a) Rat [Prescott et al.; 2009] **(b)** Shrewbot [Mitchinson et al.; 2014]

Figure 1.1 – Example for a bionic approach: from animal **(a)** to robot **(b)**.

There are definitions of each of these terms like in [Vogel; 1998], [Nachtigall; 2002], [Iouguina; 2012], [Witte et al.; 2013] or [ISO 18458] and consequently differences in the meaning and scope, but, in general, each one refers to any kind of transfer form nature to engineering. Considering the differences in these definitions, the present work aligns most with the focus of biomechatronics. There are several bionic approaches, adapting the somatosensory system of mammals, including the sensory hairs of, e.g., rats. These hairs are named vibrissae, whiskers among others, see Chapter 2. The present work stays with the term *vibrissa* in the following. Vibrissae are investigated by research laboratories all over the world. The vibrissae are arranged around the muzzle of the animals, see Figure 1.1(a). The hair shafts do not include any sensory elements. All sensory components are located inside the supporting hair follicle. Thus, the entities of vibrissae hair shafts and hair follicles are used as tactile sensors for the exploration of an unknown environment or of an object of interest. Besides further properties, they exhibit a remarkable performance in different tactile sensing tasks including the determination of a distance to an obstacle, recognizing the contour and texture of an object, or non-tactile tasks like the sensing of air flows. In Figure 1.2, the grain size of fine sand, of the tip of a stylus instrument in accordance to the profile method [ISO 3274] and the tip of a vibrissa hair shaft are compared. The example of the grain of fine sand is used to get an idea of the dimensions of the other examples. This kind of stylus instrument is used to determine, e.g., the roughness of a surface. The diameter and angle of the stylus tip are important properties of the sensor system and they determine the size of detectable surface texture elements. Since the size of a tip of a vibrissa hair shaft is of the same order of magnitude as the one of a stylus instrument, it is assumable that also the scanning performance is just as well. This reasoning as an example and the mentioned multi-functionality of the vibrissae explain the interest of engineers.

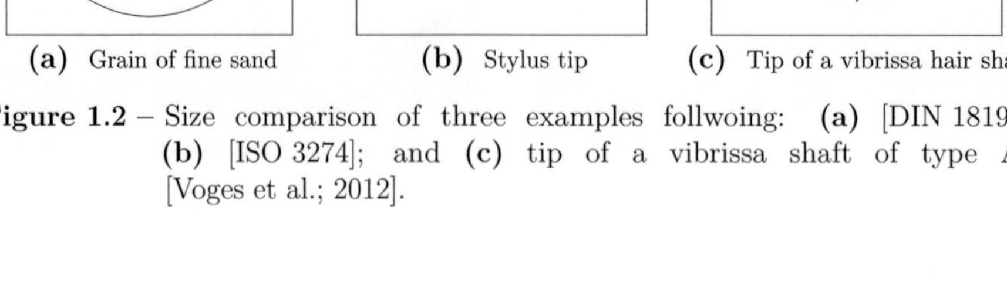

(a) Grain of fine sand **(b)** Stylus tip **(c)** Tip of a vibrissa hair shaft

Figure 1.2 – Size comparison of three examples follwoing: **(a)** [DIN 18196]; **(b)** [ISO 3274]; and **(c)** tip of a vibrissa shaft of type A2 [Voges et al.; 2012].

Starting in the late 20th century, there are plenty of works analyzing and adapting this example. From ideas adapting only the flexible hair shaft as collision detector [Hirose et al.; 1989] over works including details of the hair follicle [Mitchinson et al.; 2004] to investigations on neuronal circuits of the vibrissal system, different levels of recognition of the natural paragon are conducted.
Due to the multi-functionality, different technical applications of this sensor concept are proposed. In [Schultz et al.; 2005], [Pearson et al.; 2011] the application of these sensors and complete robot platforms (see Figure 1.1(b)) for the exploration of unknown environments/terrain mapping are discussed and demonstrated. Focusing on a single object, the profile of an object is scanned and reconstructed in [Scholz, Rahn; 2004], [Birdwell et al.; 2007], [Steigenberger; 2013] or information about the object surface texture are gathered [Neimark et al.; 2003], [Wegiriya et al.; 2019]. Also, applications to measure flow velocities and directions as well as the viscosity of fluids are reported, [Rooney et al.; 2015], [Scharff et al.; 2019b]. All mentioned scenarios can be unified since the sensor concept does not change. Indispensable for all these applications is that effects and mechanisms which are driving and influencing the sensing process must be known. There are two major ways to handle this fact regarding sensors inspired by the vibrissae of animals. First, this problem can be solved using artificial intelligence. Here, the mechanical structure of the natural paragon is adapted as well as parts of the peripheral-neuronal network. By training the network performing a finite number of training measurements it is possible to execute the mentioned tasks in a good manner. But, the inherent mechanisms and effects of the sensor system are still obscure. This point is clarified by the second possibility to handle this sensor system. The second way takes the mechanics of the sensor system into account. Mechanical modeling and analysis of the vibrissae allow for the characterization of the inherent effects and mechanisms and the influence of these on the relation between gathered information and captured signals.
The present work continues on the second way to investigate underlying mechanisms and effects while scanning an object of interest with an artificial sensor inspired by the vibrissae of rats. Following, the natural example is analyzed, based on current findings in literature to determine important structural properties as well as characteristics of the scanning process / behavior of rats during scanning, see Chapter 2. Based on this review, the biological paragon is systematically classified by the concept of a biomechatronic system [Witte, Schilling; 2017]. In this way, the unknown effects, interactions, and characteristics regarding the vibrissal system are identified and summarized in a general objective, as well as several working hypotheses, see Chapter 3. The concept of a biomechatronic system gets supplemented by the idea of a stimulus leading apparatus [Carl et al.; 2008]. These systematic analyses are necessary in order to model and design the natural example in a theoretical simulation and experiment in a way that the complexity is reduced but the important characteristics are still included, see Chapter 4. Using the theoretical model and the experiment, the inherent characteristics of the sensor systems are analyzed, see Chapter 5.

2 State of the art

The somatosensory system of mammals includes, i.a., the sense of touch. Tactile perception is investigated regarding different natural examples, see [Diamond; 2010], [Grant et al.; 2014]. One example is the human fingertip, which is involved in, e.g., grasping and nearly all tactile interactions with the environment. A fingertip consists of excitable tissue and it collects and transmits stimuli to the central nervous system [Deutsch, Deutsch; 1993]. In contrast, the sensory hair shafts of rats lack any sensory innervation [Carvell, Simons; 2017]. The innervation takes place inside the hair follicle [Ebara et al.; 2017]. Both natural systems are part of the somatosensory system but, at the same time, they are very different [Diamond; 2010]. Consequently, they are not completely comparable, and information is not exchangeable without doubts. This work focuses on the sensory hair of rats and in parts the connected periphery.

Research groups belonging to investigations with regard to the somatosensory system/the sensory hairs of rats are multidisciplinary, a selection of those is shown in Figure 2.1. They connect the areas zoology, neurobiology, taxonomy, physiology, comparative psychology, signal processing, metrology, mechatronics, and mechanical engineering among others, e.g. [Witte et al.; 2004], [Grant et al.; 2014]. In doing so, each group still focuses on particular aspects. For the present work, the mentioned research areas are of different importance. Therefore, the most known groups and their related research activities are presented in order to extract important information for this work. This review is not a complete overview of the present state of the art, in contrast, it is a selective summary to identify and explain the objective of the present work, compare Chapter 3. So mainly, the structural properties and functionalities of the vibrissal system are reviewed. Furthermore, much information about the mechanisms behind tactile sensing of rats is found based on the analyses of artificial systems inspired by natural paragon.

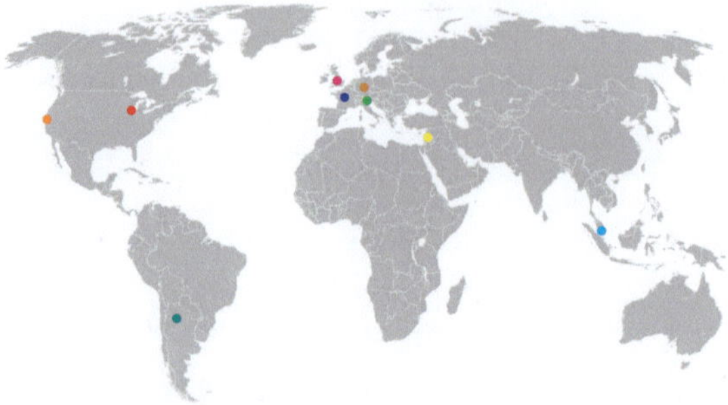

Figure 2.1 – Selection of known and active research groups belonging to vibrissae: • The Neuroscience and Robotics Lab (NxR), Northwestern University, Evanston, USA – *Mitra J.Z. Hartmann et al.* • Laboratory Jean Perrin (LJP), Sorbonne-Université, Paris, France – *Alexis Prevost et al.* • Bristol Robotics Laboratory (BRL), University of the West of England, University of Bristol, Bristol, UK – *Martin Pearson et al.* & Adaptive Behaviour Research Group (ABRG), University of Sheffield, Sheffield, UK – *Tony J. Prescott et al.* & Modelling the Mechanics of Animal Whiskers (MMEAW), London South Bank University, Manchester Metropolitan University, UK – *V. Geoffrey A. Goss, Robyn Grant et al.* • Tactile Perception and Learning Lab, International School for Advanced Studies, Trieste, Italy – *Mathew E. Diamond et al.* • Ahissar Lab, Weizmann Institute of Science, Tel Aviv, Israel – *Ehud Ahissar et al.* • Feldman Lab, University of California, Berkeley, USA – *Daniel Feldman et al.* • Bio-inspired Robotics & Design Laboratory (BRD), Singapore University of Technology and Design, Singapore – *Valdivia y Alvarado et al.* • Werner Reichardt Centre for Integrative Neuroscience (CIN), Eberhard Karls University of Tübingen, Tübingen, Germany – *Cornelius Schwarz et al.* & Bernstein Center for Computational Neuroscience Berlin (bccn), Larkum Lab, Humboldt University of Berlin, Berlin, Germany – *Michael Brecht, Robert Sachdev et al.* & Marine Science Center, University of Rostock, Rostock, Germany – *Guido Dehnhardt et al.* • Laboratorio de Medios e Interfases (LAMEIN), National University of Tucuman, Tucuman, Argentina – *Fernando D. Farfán et al.*

In the following literature review this *Reverse Bionics* and related artificial systems are not discussed in detail because the goal of the present work is to understand „Why a vibrissa is such a good tactile sensor?" and not yet „How to design a new high-performance tactile sensor inspired by natural vibrissa?" Therefore, also classic surface metrology and sensors are little included in the discussion. Summarizing, the present section must answer the following questions:

- What about the arrangement of an animal's vibrissa and its connected periphery?
- How do animals process the mechanical stimuli, and which information is included?
- How do animals behave during a particular scanning task, e.g., movement?
- How do animals decode information out of the recorded signals?

This is necessary to provide all information to set up a biomechatronic system following [Witte, Schilling; 2017].

2.1 Natural vibrissae

There are different species of mammals exhibiting sensory hairs, see Figure 2.2. All following information is related to the sensory hairs of rats and mice (see Figure 1.1(a)) because these examples are the best investigated ones in animals. Besides the information is consistent, because other species, e.g., seals, exhibit sensory hairs as well, with alike functionality but different shape.

(a) Cat **(b)** Dog **(c)** Seal

Figure 2.2 – Three further examples of animals with mystacial vibrissae: **(a)** cat of my host family in Lima, Peru; **(b)** boston terrier called Lola; and **(c)** sea lion at Islas Ballestas, Paracas, Peru.

Most topics regarding sensing with sensory hairs are related in different magnitude to neurobiology.

> The neurobiology is of particular importance for the complete process of actuation and sensing, but in the scope of the present work the *mechanical* part of the sensory hairs is in focus. So, information about neurobiology is minimized and only mentioned if necessary and useful for discussion, but there is no general review.

2.1.1 Distribution and properties

Rats exhibit sensory hairs at various locations on their body, e.g., at the paws or around the muzzle [Klauer; 2001]. The most prominent sensory hairs are located around the muzzle. These hairs are called sensory hairs, tactile hairs, sinus hairs, whisker or mystacial vibrissae [Schmidt et al.; 2014]. There is no common, conclusive definition or classification scheme that distinguishes the mentioned names in terms of, e.g., different types of hairs or functional architecture. Also, it is not defined, if these terms refer to the hair shaft or the combination of hair shaft and connected hair follicle. Sometimes, the combination of the vibrissa hair shaft and hair follicle is called follicle-sinus complex whereas others use this term exclusively for the hair follicle [Klauer; 2001]. From now on, in the present work the following conventions are used to define the system boundaries:

- The term mystacial is dropped because only mystacial vibrissae are reviewed.
- The hair shaft is called vibrissa hair shaft.
- The hair follicle is called follicle-sinus complex (FSC).
- The combination of vibrissa hair shaft and FSC is called vibrissa.

In context to structural and mechanical properties of vibrissae hair shafts, the works [Belli et al.; 2017], [Belli et al.; 2018] and [Yang et al.; 2019] of the research group around Mitra J. Z. Hartmann (see Figure 2.1) must be highlighted. They include and combine several other ones, e.g., [Voges et al.; 2012], [Carl et al.; 2012] or [Towal et al.; 2011], and they are consequently the most complete ones, the following paragraph summarizes these works and reduces the given information to the ones that are necessary for the present work.

Distribution:
Vibrissae are arranged in a particular pattern on both sides of the animal's muzzle. Following [Belli et al.; 2017], vibrissae are divided into three subcategories. The ones out of the first subcategory are located in most caudal positions. This column is denoted as Greek arc whereby the hairs are named straddlers and their position is given by the Greek letters: α, β, γ, δ from the dorsal to ventral. The rest of the array is described by rows and columns (sometimes mentioned as arcs [Voges et al.; 2012]) as well and the corresponding hairs are named macro vibrissae. There are five rows indicated by Latin letters from A - E, dorsal to ventral and the columns are numbered from 1 - 7, caudal to rostral. In further rostral direction, close to the lips, there is another type of vibrissae called micro vibrissae. In [Brecht et al.; 1997], the arrangement of the micro vibrissae is also given by a row index (F - J, dorsal to ventral). Normally, there are four straddlers, a large number of macro vibrissae, and a huge number of micro vibrissae. The exact number depends on the conditions of the examined animal, and on the used definition of the vibrissa array which can be different from the mentioned one.

In [Belli et al.; 2018] the base point position of each vibrissa hair shaft is described in 3D-space. Based on these findings, it is possible to calculate a vibrissa base position using the row and column index of the array.

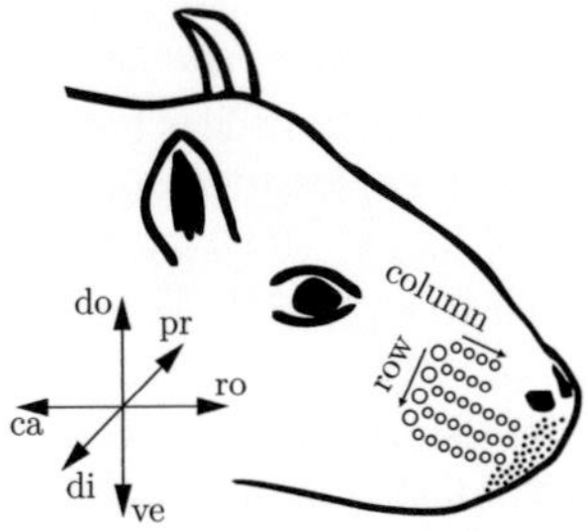

(a) Mystacial pad

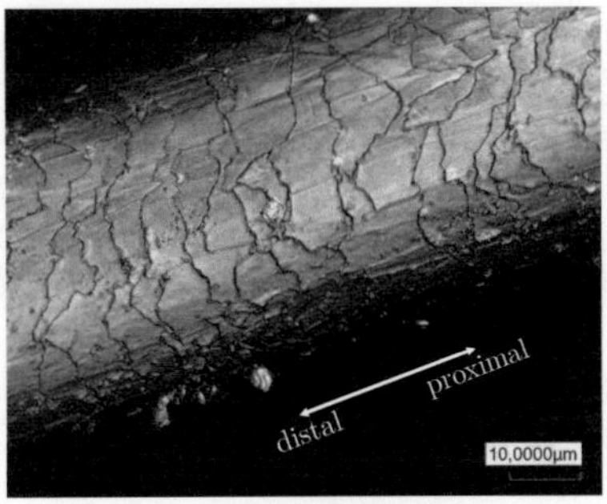

(b) Scales

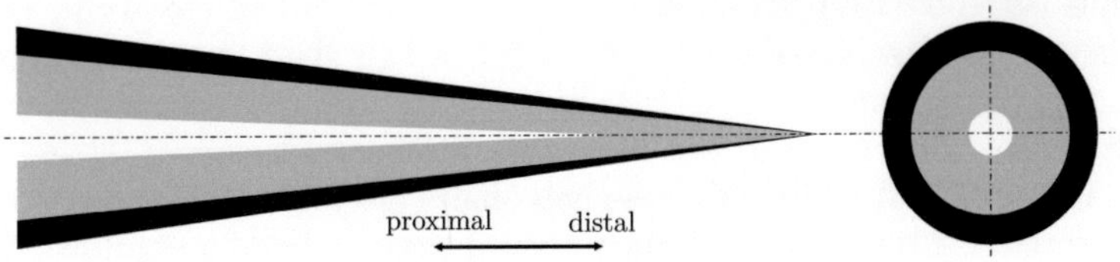

(c) Vibrissa hair shaft

Figure 2.3 – **(a)** The abbreviations of the coordinate system are: do- dorsal; pr- proximal; ro- rostral; ve- ventral; di- distal; ca- caudal. The distribution of the different types of vibrissae across the array is shown. The large circle corresponds to the straddlers: $\alpha, \beta, \gamma, \delta$ in dorsoventral direction. The smaller circles belong to the macro vibrissae. The rows are indexed from A to E in dorsoventral direction and the columns from 1 to 7 from caudal to rostral direction, following [Belli et al.; 2017]. The black dots indicate unorganized micro vibrissae. **(b)** Surface of a vibrissa hair shaft, taken from data belonging to [Voges et al.; 2012]. **(c)** Internal structure of a vibrissa hair shaft: black area- cuticle; grey area- cortex; white area- medulla. In order to illustrate a vibrissa hair shaft, the geometrical dimensions are changed and the inherent curvature is neglected. Also, the part of each layer of the total area does not correspond to the real structure. The figure is adapted from [Voges et al.; 2012].

Properties:
All types of vibrissae hair shafts share a common shape and structure. To the author's knowledge, the following information may be in parts the same for micro vibrissae hair shafts. However, there is no representative study on the morphology of micro vibrissae hair shafts because of its size and issues in classification, e.g., explained in [Grant et al.; 2012]. Straddlers and macro vibrissae hair shafts are slender and the cross section is cylindrical [Voges et al.; 2012]. Their length differs between approximately $2.9 - 66.3$ mm with an exponential decrease of the length from caudal to rostral across the array [Belli et al.; 2018]. Furthermore, longer vibrissae hair shafts correspond to larger base diameters $0.05 - 0.25$ mm [Belli et al.; 2017]. From the base of the hair shaft to its tip (proximal to distal), the diameter of the cross section is decreasing resulting in a tapered body with a base to tip diameter ratio of $4 - 57$ [Belli et al.; 2017]. In [Belli et al.; 2017] the diameter change is assumed to be linear but the results of [Hires et al.; 2016] postulate that there is a nonlinear change of the diameter. The authors of [Belli et al.; 2017] described the mentioned properties as functions based on the position of the vibrissa hair shaft in the array. Therefore, e.g., the length and the base diameter of a vibrissa hair shaft are given as functional dependence of the row and column index of the array. Another property of a vibrissa hair shaft is that the whole vibrissa hair shaft is inherently curved (stress free). It is reported that 60-70% of the vibrissae hair shafts are inherently curved in a plane [Quist, Hartmann; 2012]. The inherently curved shape follows a quadratic function $y(x) = a \cdot x^2$ whereby x and y are the Cartesian coordinates in the plane. The factor a scales the level of curvature [Knutsen et al.; 2008]. The magnitude of a depends on the length of the vibrissa hair shaft. Furthermore, a is constrained in the way that the maximum value of the y coordinate does not reach the limit of 53.3% of the corresponding vibrissa hair shaft length [Belli et al.; 2018]. A different description of the inherent curvature is reported in [Starostin et al.; 2020]. The authors found that the inherent curvature follows the trend of an Euler spiral regarding a linear function of the curvature $\kappa(s) = a \cdot s + b$ with the arc length as an argument and a, b as coefficients. Dependent on its location in the array, a vibrissa hair shaft emerges out of the skin in a particular direction. This direction is described by three angles and can be calculated considering row and column index [Yang et al.; 2019]. Doing so, the authors of [Yang et al.; 2019] demonstrate that there are regularities across the array, e.g., a change from ventral to dorsal.

A vibrissa hair shaft consists of dead cells [Rice; 1993] that form three layers of material. The outside/outermost layer of the hair shaft, the cuticle, is made of approximately 15 - 20 overlapping coatings of scales with an average diameter of 5.43 μm [Yanli et al.; 1998], [Voges et al.; 2012]. The cuticle and the cortex (second layer) are present along the whole length of the hair shaft. The cuticle belongs to an average amount of 16% of the cross section consisting of β-keratin, while the cortex belongs to 67% - 84% consisting of α-keratin [Voges et al.; 2012], [Yang et al.; 2019]. The cortex encases the medulla.

The medulla is the innermost part of a vibrissa shaft, it is conical and hollow [Voges et al.; 2012]. Across the array of vibrissa, the medulla changes its length from 50% - 75% of the total vibrissa shaft length. This characteristic is also reported for the diameter of the medulla at the base of the hair shaft which changes between 6% and 15%. The density of a vibrissa hair shaft is changing along its length, it has an average density of $\rho = 1280 \pm 160\,\mathrm{kg\,m^{-3}}$ [Yang et al.; 2019]. The change of the density is caused by the changing amount of the different material layers. A similar characteristic is reported for the Young's modulus, there are smaller values at the base and larger ones at the tip [Quist et al.; 2011]. The Young's modulus of the cuticle is larger than the one of the cortex which can be related to the different materials of the cuticle and cortex. Average values of Young's modulus for a complete vibrissa hair shaft are reported in a range from $3.3\mathrm{e}8\,\mathrm{N\,m^{-2}}$ [Kan et al.; 2013] to $7.36\mathrm{e}9\,\mathrm{N\,m^{-2}}$ [Carl et al.; 2012]. In [Huet et al.; 2017] the shear modulus is defined as for an isotropic, linear elastic material regarding Young's modulus and Poisson's ratio. Following [Huet et al.; 2017], [Kan et al.; 2013], the Poisson's ratio differs from 0.38 to 0.4. Dynamical investigations on isolated vibrissae hair shafts show that the natural frequencies of the different vibrissae hair shafts increases from caudal/dorsal to rostral/ventral direction in a range of $26-750\,\mathrm{Hz}$ [Neimark et al.; 2003], [Yan et al.; 2016] and the damping ratios from 0.11 to 0.17 [Hartmann et al.; 2003]. Further mechanical properties of vibrissae hair shafts are evaluated in [Yang et al.; 2019]. The center of mass, the moment of inertia, and the second moment of area get smaller from dorsal to ventral direction across the complete array.
Every vibrissa hair shaft is supported by an FSC [Vincent; 1913]. The vibrissa hair shaft and its FSC are fixed to each other [Bagdasarian et al.; 2013]. The FSC and the surrounding elastic tissue encapsulate the base of the vibrissa hair shaft along a length of $3-5\,\mathrm{mm}$ [Ebara et al.; 2002]. An FSC consists of various types of tissues and blood vessels, see Figure 2.4. The whole assembly of capsule of the FSC including the components in Figure 2.4 and the base of a vibrissa hair shaft is relatively rigid. The mechanical properties of an FSC are widely unknown. The authors of [Mitchinson et al.; 2004] summarizes and estimate the masses, density, spring constants, and critical damping coefficients for different components of an FSC. These values are more optimized to fit a theoretical model and less a representation of the real components. For example, the spring constants are estimated considering a Young's modulus $1\mathrm{e}7\,\mathrm{N\,m^{-2}}$of rat skin. An alike parameter summary for a row of vibrissae inside the array is given in, e.g., [Simony et al.; 2010]. Besides structural components, an FSC is innervated by thousands of mechanoreceptors [Szwed et al.; 2003]. They are distributed along the part of the vibrissa hair shaft that is encapsulated by the FSC [Ebara et al.; 2017], see Figure 2.4. The mechanoreceptors and the connected periphery transform a stimulus into an action potential (electrical signal). For each FSC, the action potential is transmitted by one deep vibrissal nerve (innervation of the lower part of an FSC) and multiple superficial vibrissal nerves (innervation of the upper part of an FSC), both are part of the trigeminal nerve (afferent/sensory nerve).

Particular types of mechanoreceptors belong to either the deep vibrissal nerve or the superficial vibrissal nerve [Ebara et al.; 2017]. Generally, mechanoreceptors can be grouped into two categories, there are slow adapting (SA) and rapid adapting (RA) ones. Following [Soderquist; 2002], simplified, to a stationary/quasi-static stimulus:

- RA mechanoreceptors respond initially with a very high-frequency sequence of action potentials, then adaption takes place and the frequency of the action potentials get very low;
- SA mechanoreceptors respond initially with a high-frequency sequence of action potentials, then adaption takes place and there is a continuous sequence of action potentials with a moderate frequency.

In case of a fast oscillating stimulus, the RA mechanoreceptors will respond adequately since they are sensitive to change in a time signal [Soderquist; 2002]. Both types of mechanoreceptors have an activation threshold. Furthermore, a more intensive stimulus will increase the frequency of the action potential but not its amplitude [Soderquist; 2002], [Deutsch, Deutsch; 1993]. Without getting deep into neurobiology, the signals recorded by the mechanoreceptors are processed various times before reaching the central nervous system, e.g., in the primary sensory neuron. Even if neurobiological aspects of the vibrissal system are out of the scope of the present work and mostly neglected, they are an important part for sensory systems.
A complete FSC with all its components is protected by a stiff capsular [Ebara et al.; 2002] and is pivoted at its upper end [Kim et al.; 2012], see Figure 2.5. The base of a vibrissa hair shaft and the FSC are moved together, like a single rigid body by different muscles [Bush et al.; 2016]. The muscles that are connected to vibrissae are distinguished in intrinsic and extrinsic muscles [Dörfl; 1982]. Muscles that are classified as intrinsic, originate and end between the FSCs in the array whereby the origins of the extrinsic ones are beyond the region of the array [Dörfl; 1982], [Haidarliu; 2015]. There are three types of intrinsic and five types of extrinsic muscles [Haidarliu; 2015], [Haidarliu et al.; 2017]. The muscles and surrounding skin (visco-elastic tissue) allow for a rotation of a single vibrissa in rostrocaudal and dorsalventral direction, furthermore, a vibrissa shaft gets rotated with respect to its long axis [Haidarliu et al.; 2017]. These findings are restricted due to the fact that the intrinsic muscles connect always two adjacent vibrissae, so if one is moved always a second one is moved as well but with a different amplitude [Haidarliu; 2015]. Also, the complete array can be moved, all vibrissae can be rotated or translationally moved together in rostrocaudal direction [Haidarliu; 2015]. The spread between the vibrissae is controllable as well [Haidarliu; 2015]. Besides, the muscles, also the surrounding skin contributes to the movement of the vibrissae by its visco-elastic properties [Haidarliu; 2015]. All mentioned muscles are innervated by the facial nerve (efferent/motor control) [Haidarliu; 2015]. The facial nerve in combination with the trigeminal nerve represents a closed-loop motor sensory control [Simony et al.; 2010] that enables advanced sensing strategies. All findings regarding the musculature refer to straddlers and macro vibrissae, micro vibrissae are not actuated by particular muscles [Prescott et al.; 2011], [Grant et al.; 2012].

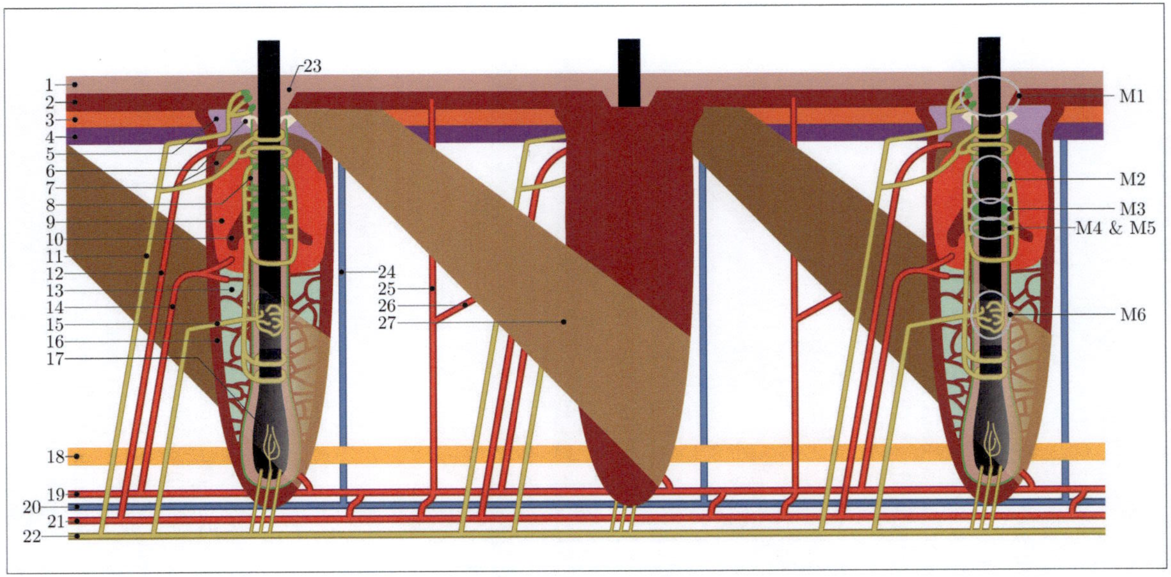

Figure 2.4 – Left, mid FSCs adapted from [Bosman et al.; 2011],[Haidarliu; 2015]: 1- epidermis (rose); 2- dermis (dark red); 3- m. nasolabialis (orange); 4- m. maxillolabialis (magenta); 5- outer conical body (light magenta); 6- sebaceous gland (beige); 7- inner conical body (brown); 8- glassy membrane (green); 9- ring sinus (red); 10- ring wulst (dark red); 11- superficial vibrissal nerves (dark yellow); 12- superficial vibrissal arteries (red); 13- cavernous sinus (cyan/red); 14- deep vibrissal arteries (red); 15- deep vibrissal nerve (dark yellow); 16- capsule (dark red); 17- vibrissa hair shaft (black); 18- m. nasolabialis profundis (light orange); 19- intervibrissal trunks (red); 20- row veins (blue); 21- row arteries (red); 22- trigeminal nerve (dark yellow); 23- rete ridge collar (rosa); 24- intervibrissal veins (blue); 25- intervibrissal arteries (red); 26- arterioles (red); 27- intrinsic muscle (brown). Right FSC adapted from [Bosman et al.; 2011], [Ebara et al.; 2017]: M1- merkel endings (SA&RA); M2- circumferential lanceolate endings (-); M3- merkel endings (SA); M4- longitudinal lanceolate endings (RA); M5- club-like endings (RA); M6- free nerve endings (-).

2.1.2 Whisking and scanning behavior

Behavior studies demonstrate that there are three situation related to sensing with vibrissae, following [Curtis, Kleinfeld; 2006] there is: a startle mode, an exploration mode, and an object detection mode.

Startle mode:

The startle mode is also known as receptive mode [Diamond, Arabzadeh; 2013] or passive mode, e.g., [Kleinfeld et al.; 2006]. Here, the vibrissae are not actuated by the connected muscles, they are in a resting position. But there can be a secondary movement like the overall locomotion or head movement of the animal. In order of these movements, a stimulus can arise due to contact with an object.

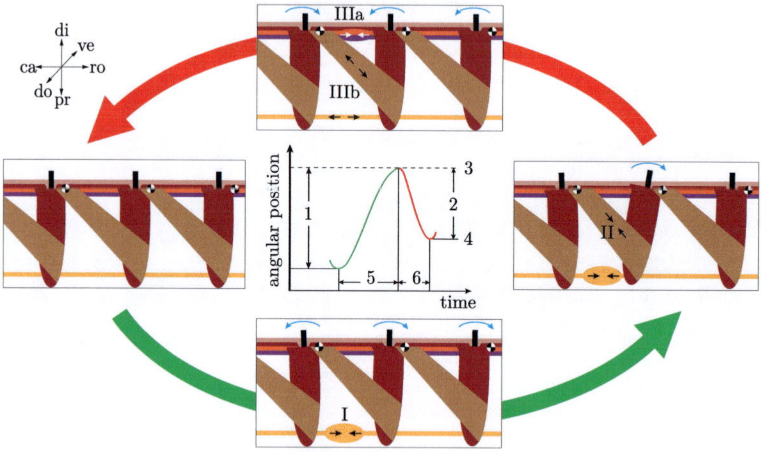

Figure 2.5 – Whisk sequence from left-bottom-right-top adapted from [Berg, Kleinfeld; 2003], [Bosman et al.; 2011], [Haidarliu; 2015]. Protraction (green arrow): left-bottom-right; retraction (red arrow): right-top-left. Left: resting position; bottom: I- contraction of m. nasolabialis profundus (extrinsic muscle); right: II- still contracted m. nasolabialis profundus and contraction of the intrinsic muscle; top: IIIa- contraction of m. nasolabialis and m. maxiolabialis (extrinsic muscles), IIIb- relaxation of m. nasolabialis profundus and instrinsic muscles. The blue arrows indicate the rotation direction of each shown vibrissa. Center, qualitative whisking parameters adapted from [Hartmann; 2011]: 1- protraction amplitude; 2- retraction amplitude; 3- protraction setpoint; 4- retraction setpoint; 5- protraction duration; 6- retraction duration. The author of the present work proposes the displacement of the adjacent vibrissae considering the findings of, e.g., [Simony et al.; 2010].

Furthermore, even if there is no movement of the animal at all, the vibrissae can be deflected by an external stimulus, e.g., a moving object. Investigations on sensing in startle mode are controversially discussed because most experiments are done with anesthetized rats in particular experimental setups and not in behavioral studies [Diamond, Arabzadeh; 2013], [Towal et al.; 2012]. So, it is still not totally clear how exactly and in which moments animals use the startle mode. One exception is reported for sensing of air flows with vibrissa [Yan et al.; 2016].

Exploration mode:
The exploration and object detection mode are frequently combined in the term active or generative mode [Diamond, Arabzadeh; 2013]. Since the vibrissae get actively actuated in this mode, the mechanoreceptors capture signals related to the induced displacement/forces of the vibrissae [Szwed et al.; 2003], [Diamond; 2010]. Other researchers use the terms non-contact whisking and contact whisking to differentiate between both situations [Hartmann; 2015].

The exploration mode is characterized by an active actuation. So, rats sweep their vibrissae rhythmically forward and backward in rostrocaudal direction free in air without any contact or collision. This movement is known as a whisk or whisking cycle [Berg, Kleinfeld; 2003]. In detail, a whisk includes three situations of actuation [Towal et al.; 2012], [Haidarliu; 2015], see Figure 2.5. First, the vibrissa shaft gets protracted slightly forward by the extrinsic muscle (m. nasolabialis profundus), see Figures 2.4 and 2.5. Following, this protraction is forced by the intrinsic muscles. Afterwards, the vibrissa shaft retracts by both the extrinsic muscles (m. nasolabialis and m. maxiolabialis) and the visco-elasticity of the surrounding tissue until it reaches the initial state again. The part of the whisk in rostral direction is called protraction and the one in caudal direction retraction. The displacement in rostrocaudal direction is superimposed by further movements. There is a movement in dorsoventral direction as well and consequently a vibrissa shaft moves along an elliptical trajectory [Bermejo et al.; 2002]. Finally, during a whisk, a vibrissa shaft gets rotated around its longitudinal axis [Knutsen et al.; 2008], [Haidarliu et al.; 2017]. In a typical whisk in exploration mode, vibrissae shafts are rotated within a frequency of $5-15\,\mathrm{Hz}$ and with amplitudes of approximately $\pm 25°$ in rostrocaudal direction [Berg, Kleinfeld; 2003], [Towal et al.; 2012] and of $\pm 7°$ in dorsoventral direction [Bermejo et al.; 2002]. The combination of frequency and amplitude results in a protraction angular velocity of $250-1000\,°\mathrm{s}^{-1}$ and in the case of retraction $250-1500\,°\mathrm{s}^{-1}$ [Towal et al.; 2012]. Rats are able to perform both symmetric and asymmetric whisking with respect to the left and right vibrissa array [Towal, Hartmann; 2006].

Object detection mode:
The object detection mode includes various sub-modes and scanning strategies and behaviors. One of these is exploratory whisking. As the name indicates, this sub-mode is closely related to the exploration mode.

It is also characterized by a whisking motion, but in contrast to the exploration mode during exploratory whisking the vibrissa hair shaft touches an object (active touch) [Hartmann et al.; 2003]. A typical whisk of exploratory whisking is characterized by a frequency of 8 Hz, an amplitude of $24 - 32°$ and an angular velocity of $696\,°\mathrm{s}^{-1}$ for protraction and $1106\,°\mathrm{s}^{-1}$ for retraction [Carvell, Simons; 1990], [Pammer et al.; 2013]. Slightly different are the findings in [Grant et al.; 2008]. The authors reported that the animals stop the protraction of the vibrissae after an initial contact immediately and retract them with a lower velocity than used as for protraction. In the same work, this behavior is related to another scanning strategy called minimal impingement. Minimal impingement means that the animals tend to minimize the amount bending of their vibrissae in consequence of a surface contact [Mitchinson et al.; 2007]. Furthermore, the minimal impingement is accompanied by a second strategy, which implies that animals aspire a maximal number of contacts between vibrissae hair shafts and surface [Mitchinson et al.; 2007]. This finding is supported by the fact that animals decrease actively the spread between their vibrissae to increase the number of contacting vibrissae [Grant et al.; 2008].

The exploratory whisking can be complemented by a dabbing motion of the head in order to dab the micro vibrissae against the object. In this case, the animals use the larger macro vibrissae to orientate their head in the right position and afterwards they start dabbing against the object [Hartmann; 2001], [Grant et al.; 2012]. The dabbing motions are performed with a frequency of 8 Hz and an amplitude of 5 mm.

Another sub-mode is the so-called foveal whisking and reported in [Berg, Kleinfeld; 2003]. Foveal whisking is characterized by a frequency of $15 - 25$ Hz and small amplitudes (there are no values reported as far as the author knows). The vibrissae are orientated forward in a way that the tips of the vibrissae hair shafts form a dense arrangement in front of the animal's snout. This forward movement is mainly driven by the intrinsic muscles [Berg, Kleinfeld; 2003], an additional activation of the extrinsic muscles to minimize the vibrissae spread is proposed in [Haidarliu; 2015]. The backward movement is done passively by the visco-elasticity of the connected tissue. Animals use foveal whisking to scan an object in detail after they detected it in exploration mode by collision.

The variety of different scanning modes and actuation strategies leads to the assumption that in case of an active movement of the vibrissae, the motor signal is important for the scanning process [Ganguly, Kleinfeld; 2004], [Diamond, Arabzadeh; 2013] and that every task requires a particular mode and/or strategy. But, in freely moving rats different motions, like head movement as well as breathing, sniffing, and whisking are superimposed [Hartmann; 2011], [Kleinfeld et al.; 2015]. Consequently, the closed-loop sensory-motor control is adaptive in order to react to different scanning scenarios and includes a central pattern generator for the coordination of different mechanisms underlying rhythmic whisking [Ahissar, Knutsen; 2011], [Kleinfeld et al.; 2015].

2.1.3 Functionalities and signal processing

The functionalities of vibrissae include very different tasks, see Figure 2.6. For example, vibrissae are important for social behavior, food acquisition/hunting, selecting nest material, locomotion, navigation, detection of air flows, and various task of tactile discrimination of objects of interest among others [Ahl; 1986], [Evans et al.; 2019], [Diamond, Arabzadeh; 2013]. For the present work, the mentioned functions are of different importance, e.g., the influence on the social behavior of rats is less important than the capabilities of tactile discrimination.

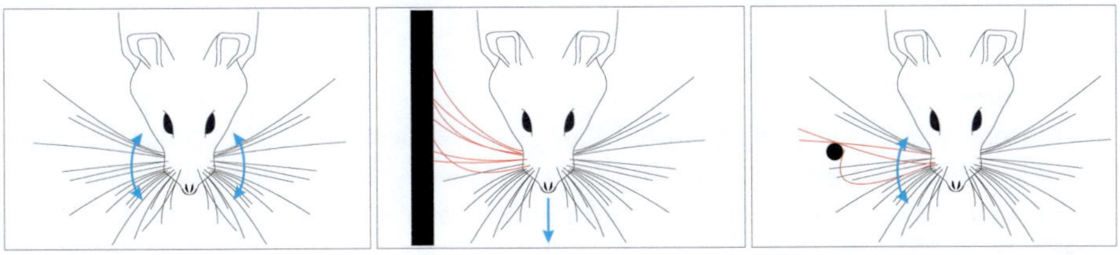

(a) free whisking (b) wall following (c) object localization

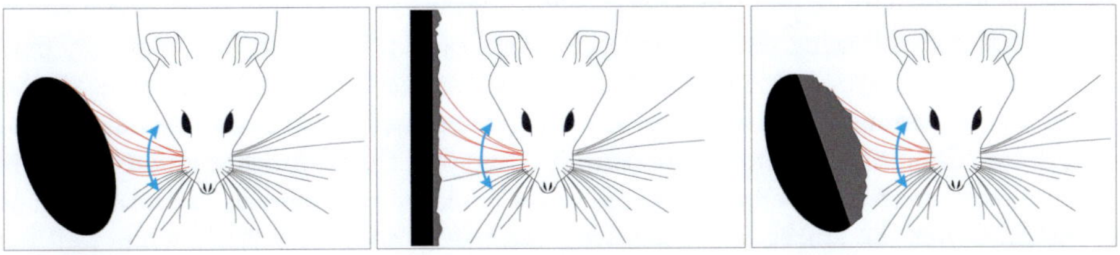

(d) shape recognition (e) texture detection (f) combined task

Figure 2.6 – Illustration of different scanning tasks. Deflected vibrissae are colored in red and undeflected ones in black. The blue arrows indicate the direction of movement.

The mentioned examples are not independent of each other, so the selection of nest material depends on the discrimination of the object surface texture [Diamond, Arabzadeh; 2013] and so on.
Therefore, only tasks related to tactile discrimination of objects are reviewed in the present work. Namely, distance estimation, wall following, object contour recognition, and surface texture detection. Between these, there is a difference in the kind of information that is captured. Firstly, information out of wall following and object localization tasks answer the question „Where is the object?“, and secondly, signals out of object contour and texture scans contain information related to „What is the object looking like?“, [Diamond et al.; 2008b], [Garrett; 2009]. Independent from the actual sensing task, there are some general remarks. All mentioned tasks are related to the position of the complete animal, its head, and each vibrissa in any wise.

So, different studies use different reference frames and types of coordinate systems but they are related to each other [Ahissar, Knutsen; 2011], [Hartmann; 2015]. Summarizing, there are two important reference points: the head and the FSC. These points are used to introduce the head centered coordinates and vibrissa centered or vibrissa resting coordinates. Consequently, the signals are captured in a reference system related to the orientation of the FSC and by further processing transformed into spatial information related to the head of the animal [Hartmann; 2011]. This seems to be possible because the current vibrissa orientation is probably recorded by the mechanoreceptors [Szwed et al.; 2003].

Wall following:
In [Vincent; 1912] it is firstly reported that rats use their vibrissae to follow the walls of a labyrinth, see Figure 2.6(b). In the experiments, the rats were blindfolded. In the first turn, rats had all vibrissae and moved safely through the labyrinth. In the second turn, the vibrissae were clipped, and the rats collided often with walls and edges. So, there is evidence that rats use their vibrissae to guide locomotion based on stationary stimuli and that the location of the wall is tracked with reference to their heads. The authors of [Sofroniew et al.; 2015] assumed that rats track the contact force while wall following. Additionally, in [Jenks et al.; 2010], the low frequency part of the captured signal is identified to contain information about the distance to the walls. The experiment in [Kerekes et al.; 2017] demonstrates that rats can follow walls and detect details of the surface texture by dragging their vibrissae along the walls while moving with a velocity of approx. $1\,\mathrm{m\,s^{-1}}$. Also, in [Kleinfeld et al.; 2006] it is mentioned that the vibrissae are passively dragged across the surface while wall following. This functionality is implemented in numerous artificial sensor concepts, e.g., [Schroeder, Hartmann; 2012]. The authors of [Schroeder, Hartmann; 2012] propose that rats can predict future contact points based and the captured signals (forces and moments) at the base of the vibrissa hair shaft in order to optimize their locomotion.

> These studies show that the rat moves along a given trajectory whereby multiple vibrissae contact the wall in a passive manner, there is no whisking. Furthermore, the signals seem to be continuously processed for a longer duration of time.

Object localization:
This task is closely related to contact/collision detection but with an important difference. In case of contact/collision detection, there is one information if there is or if there is not a contact/collision. Whereas object localization also includes information where did the contact/collision happen, see Figure 2.6(c). This statement is not a common fixed definition, it is just to distinguish between the mentioned cases. Across vibrissae related literature, there are multiple proposed coding schemes for an object localization procedure.

The authors of [Ahissar, Knutsen; 2011] identify four coding dimensions: space, intensity, time, and posture. These dimensions are combined into six procedures to determine the object location [Cheung et al.; 2019]. In parts, the procedures use very easy concepts like the determination of the distance to an obstacle based on the position of the vibrissae in the array [Brecht et al.; 1997]. The length of the vibrissae hair shafts is varying systematically across the array, see Section 2.1.1. Thus, if the first vibrissa does not contact the object but the following longer vibrissa does, the object is at least placed in a distance that is larger than the length of the first vibrissa. In contrast, there are also very complete procedures like in [Knutsen, Ahissar; 2009] that include various different coding dimensions and not only the vibrissa identity. All these procedures propose task specific mechanisms but considering that the animals can perform different tasks with the same sensor a more general mechanism seems to be reasonable. There are a lot of works demonstrating that mechanical quantities at the base/inside the FSC in consequence of contact between vibrissa hair shaft and object are sufficient to determine the object location in different scenarios. In [Campagner et al.; 2018] these mechanical quantities are divided into two groups: kinematic variables (vibrissa position and velocity) and mechanical variables (forces and moments). The authors review studies over the last decades and summarize that forces and moments inside the FSC are the best way to determine the distance to an object, this is also consistent with findings regarding wall following as mentioned before.

From here on, these reactions forces and moments are called support reactions.

The support reactions are represented as a time series [Hartmann; 2015]. Tracking the shape of vibrissae hair shafts of head-fixed mice during object contact, the authors of [Pammer et al.; 2013] demonstrate that by calculating the support reactions corresponding to the deformed shapes, the object location can be determined. Furthermore, the object location was recognized with one-millimeter accuracy and only one vibrissa was used. These findings encounter hypotheses to judge the object location based on vibrissa identity. Using a theoretical model that respects typical properties of vibrissae, e.g., a tapered shape or inherent curvature, the authors of [Huet et al.; 2017] confirm these findings. Here, it is shown that an object location in 3D-space is uniquely coded by forces and moments at the base of a vibrissae hair shaft. However, it is important to point out that this finding is limited and constrained by the underlying mechanical theory and related assumptions.
Determining the object location based on knowledge of the support reactions includes two different scenarios. Both, the signals as result of the impact between the vibrissa hair shaft and object [Boubenec et al.; 2012], [Baldeweg et al.; 2014], [Claverie et al.; 2017], [Vaxenburg et al.; 2018] as well as the support reactions as result of the bending deformation [Birdwell et al.; 2007], [Pammer et al.; 2013], [Quist et al.; 2014], [Huet et al.; 2017] can be used.

Contacting an object, both cases will occur simultaneously, so the impact causes an high-frequency vibration which is recorded by RA mechanoreceptors, while the bending deflection is related to a low-frequency signal captured by SA mechanoreceptors [Claverie et al.; 2017], [Vaxenburg et al.; 2018].
Which of both procedures is actually used by animals is not known. It is not even known if, e.g., the bending moment or the change of the bending moment at the base of the vibrissa hair shaft is used to determine the object location. Most probably animals use a combination out of different mechanisms [Solomon, Hartmann; 2011].

In this way, it can be assumed that the object distance is determined by evaluating the high-frequency part of the signal to avoid collisions and to get a fast feedback about the distance. Furthermore, evaluating the low-frequency part of the signal is used to detect further detailed object features.

Object shape recognition:
A frequently mentioned functionality of the vibrissal sensory system is the detection of the contour of an object, see Figure 2.6(d). There are studies demonstrating that rats and mice can recognize the shape of objects. In [Brecht et al.; 1997] a rat had to identify the tasty cookie by evaluating its shape, for this case it is reported that the rat used the micro vibrissae to do so. In contrast, the authors of [Harvey et al.; 2001] found that a rat with a fixed head can distinguish/recognize a spherical object from a square one by using its macro vibrissae only. An alike experiment for freely moving mice is done in [Hu et al.; 2018]. Here, mice were able to differentiate between a square and a triangular-shaped object. All three exemplarily chosen works show that animals can recognize the shape of an object. But, due to issues in observing the behavior of freely moving animals while scanning objects, it is still not exactly explained how do animals recognize object shapes [Hartmann; 2011]. The vacancy is mainly filled out by artificial vibrissa-like sensors adapting the paragon of a macro vibrissa [Diamond et al.; 2008b]. Like in an object localization task, there are different ways to detect an object shape. In different works, e.g., [Kim, Möller; 2007], [Solomon, Hartmann; 2011], the authors propose procedures based on the identity and corresponding length of each vibrissa or based on the support reactions in consequence of a contact among others. In [Scholz, Rahn; 2004], [Clements, Rahn; 2005] an approach that uses only the support reactions as a source of information about the object shape is presented. This method includes a theoretical framework that allows for a recognition of 3D-shapes, but it is only demonstrated for a 2D-shape. This approach considers various findings regarding the whisking of animals, e.g., the sensor is rotated against the object and, continuously in time, swept across its shape like in exploration whisking. The signals are reported at the base of the sensor shaft as in an FSC and the sensor shaft is highly elastic alike to a vibrissa hair shaft, therefore the framework includes also large deformations of the sensor shaft.

A complete reconstruction of a 3D-object using a vibrissa-like sensor array is demonstrated in [Schultz et al.; 2005], [Solomon, Hartmann; 2006]. In difference to the first mentioned approach, an array of vibrissae is used. The underlying mechanical theory covers only small deflections/deformations and the sensors shafts only dab against the surface to limit the amount of deflection. These reversal findings, on the one hand, large deflections and on the other hand small deflections but in both cases an accurate sensing result, reflects the findings regarding natural whisking/sensing [Hartmann et al.; 2003], [Mitchinson et al.; 2007], [Grant et al.; 2008]. Since one of the most important properties of a vibrissa is that the vibrissa hair shaft is highly elastic, flexible, and suffers large deformations without getting damaged (compare Subsection 2.1.4), it is necessary to include this characteristic in the investigations. Both, wall following and an evaluation of an object shape with vibrissa-like sensors are analyzed by the authors of [Schroeder, Hartmann; 2012]. They show that both tasks are alike and can be performed using knowledge about the current support reactions. The used mechanical model is strongly constrained by underlying assumptions, e.g., a contact between the sensor shaft and the object can only arise along the sensor shaft and not at its tip. This issue is considered in the theoretical model of [Steigenberger; 2013]. Here, a contact along the vibrissa hair shaft as well as at its tip is covered by an appropriated mechanical and mathematical model. Furthermore, this model takes large deflections into account and allows for a rotation and translational displacements [Will et al.; 2017], [Merker et al.; 2018]. All discussed models use the low-frequency part of the support reactions, so that can be taken as a further indicator that object shape is coded in this part of the signal.

Comparing the findings regarding object shape recognition to the ones of wall following and determination of the object distance, there are similarities. While detecting an object shape, the support reactions are captured and evaluated, e.g., [Scholz, Rahn; 2004]. The characteristics of the trends of the support reactions correspond to a low-frequency signal as mentioned for wall following and one possibility of locating an object [Bush et al.; 2016]. Furthermore, in general, it is reported that a time series signal can be integrated by neurons [Soderquist; 2002]. In specific, for rats, it is reported that they accumulate information over a certain time [Fassihi et al.; 2015] and that they perform continuous whisks along an object shape [Hartmann et al.; 2003], as well as an integration over multiple touches and/or vibrissae takes place [Sofroniew, Svoboda; 2015], [Ego-Stengel et al.; 2019]. Multiple touches can be realized by contact with multiple vibrissae or by contact with one vibrissa in combination with a memory about the previous contact [Schroeder, Hartmann; 2012]. These facts lead to the assumption that the object shape can be reconstructed by a finite number of contact point locations [Bush et al.; 2016]. Combining the recorded points, a cloud of points located on the object surface results as an approximation of the real shape of the contacted object. This supposed procedure is difficult to realize if the object location is evaluated on base of the impact and corresponding high-frequency signal.

This idea supports the assumption that there are different mechanisms for different tasks. A rat, exploring the environment, can estimate the distance to unknown and unexpected obstacles to avoid collisions based on a very fast signal/procedure as well as do an extraction of detailed object properties like the shape based on a whisk with a longer time duration.

Texture detection:
The ability of rats and mice to discriminate surface textures is extensively investigated, see Figure 2.6(e). In [Guić-Robles et al.; 1989] it is demonstrated that rats rely strongly on their tactile input of the contact between macro vibrissae and object surface while differentiating a smooth from a rough surface, see Table 2.1. This behavioral study is an early demonstration of texture determination but does not give information about how the process of texture determination works. The authors of [Carvell, Simons; 1990] reported that rats reduce the whisking velocity and amplitude while touching an object (compare Subsection 2.1.2), whereby the protraction is slower than the retraction of the vibrissa hair shaft. Furthermore, it is reported that the contact between vibrissa hair shaft and surface takes place along the hair shaft and not at its tip which is controversial to the findings in [Mitchinson et al.; 2007]. Both works confirm that rats tend to use more than one vibrissa hair shaft for scanning. The investigations of [Carvell, Simons; 1990] were continued in [Carvell, Simons; 1995]. Here, the scanned surface textures are classified as macroscopic and microscopic and not as smooth and rough. It is found that the macroscopic surface can be detected by simply recognizing the presence of the larger macroscopic features. For both surfaces, the amount of bending of the vibrissa hair shaft is comparable and seems to be controlled in order to optimize the scanning conditions for capturing the signals with SA mechanoreceptors. Therefore, the authors assume that a difference in amplitude and frequency of the surface induced vibrations is used to differentiate the surfaces. Like in the previous work [Guić-Robles et al.; 1989], the rats distinguish between different surface texture and in addition, the whisking kinematics during surface texture detection are investigated but the underlying mechanisms are little discussed. A first attempt to explain the mechanisms underlying the detection of a surface texture is done in [Neimark et al.; 2003]. This work is extended in [Moore, Andermann; 2005] and hence, the authors formulate the so-called *Resonance Hypothesis.* This hypothesis is alike to the process of object localization based on vibrissa identity. Here, it is suggested that a surface texture excites a vibrissa hair shaft and the hair shaft starts oscillating. Based on the location in the array, the geometry of a vibrissa hair shaft changes and consequently, its natural frequency differs across the array, see Subsection 2.1.1. Thus, if the frequency of the induced vibration is close to the natural frequency of a vibrissa hair shaft the vibrissa hair shaft starts to oscillate with a larger amplitude than the others. Following this idea, in [Neimark et al.; 2003] smooth and rough surfaces are scanned with different vibrissa, see Table 2.1.

The results show that if a vibrissa hair shaft gets excited with a frequency close to its natural frequency the signal gets amplified. This effect leads the authors to the conclusion that both, the surface texture can be determined, and small signals get amplified to a perceptible level by the resonance of a particular vibrissa in the array.

Another hypothesis, the *Kinetic Signature Hypothesis*, was reported in [Arabzadeh et al.; 2005]. This hypothesis states that a surface texture is represented by a sequence of velocity events while the vibrissa hair shaft contacts the surface. The sequence of these events is characterized by the number, occurrence, and temporal spacing of the events. In [Diamond et al.; 2008a], [Wolfe et al.; 2008], it is mentioned that resonance of a vibrissa hair shaft is a side effect and not part of the underlying coding mechanism. This view is supported by the findings of [Simons; 1995], there, rats with only one vibrissa hair shaft (the others were clipped) successfully detected different surfaces which is unlikely for the *Resonance Hypothesis*.

Later on, the velocity events were associated with the stick-slip effect and that frequency and amplitude of slips represent the surface, see [Diamond et al.; 2008a], [Ritt et al.; 2008], [Wolfe et al.; 2008]. These events are accompanied by ringing of the vibrissae hair shafts after slips. The authors of [Wolfe et al.; 2008] show that a smooth surface corresponds to a large number of low amplitude events and a very small number of high amplitude events. In contrast, a rough surface is represented by a moderate number of low amplitude events and an increasing number of high amplitude events compared to the smooth surface. An alike observation is done in [Zuo et al.; 2011]. Here, rats discriminated grooved surfaces (see Table 2.1) using their macro vibrissae. In this work, it was found that the probability of the occurrence of stick-slip events was higher for surfaces with larger groove spacing. Hence, the authors concluded that only evaluating stick-slip events is not sufficient to detect a surface texture. Complementary, neuronal recordings in the cortex of a rat showed that the amplitude and frequency of an applied sinusoidal stimulus cannot be determined independently [Adibi et al.; 2012]. All these findings are combined in the *Slip Hypothesis* in [Schwarz; 2016]. This hypothesis favors the discrete stick-slip events and their spatial as well as temporal distribution as the underlying mechanism. Furthermore, it is highlighted that the stick-slip effect is driven by friction and not necessarily by elements of the surface. The author limits the hypothesis to microscopic scale. Observing rats while scanning a smooth and a grooved surface is found that both surface types are encoded by the stick-slip effect [Isett et al.; 2018]. In context to the *Slip Hypothesis*, in [Oladazimi et al.; 2018] is demonstrated that the signal of the acceleration of a vibrissa hair shaft is affected by the stick-slip effect, and based on this signal a texture can be detected by identifying the included stick-slip events. Attempts to evaluate other recorded signals like the velocity or using other methods to evaluate the signals, e.g., by Fourier analyzes, were outperformed by analyzing the acceleration and stick-slip events. For the experiments, different natural vibrissae hair shafts were fixed onto a motor and rotated along a circular-shaped object covered with sandpaper, see Table 2.1. Applying different scanning velocities and object distances as well as using different vibrissae hair shafts the discriminability changed.

Table 2.1 – Exemplary overview of used surface textures and related classification as smooth or rough.

	smooth	rough
[Guić-Robles et al.; 1989]	sandpaper 400μm grain size $\approx$ 200 grains/cm^2 u-shaped carrier	sandpaper 2000μm grain size $\approx$ 2 grains/cm^2 u-shaped carrier
[Carvell, Simons; 1990]	not specified plastic cylinder (axial)	30μm grooves 90μm spacing plastic cylinder (axial)
[Carvell, Simons; 1995]	microgeometric 5μm deep & 15μm wide grooves spacing 50-500μm plastic cylinder (axial)	macrogeometric 1000μm grooves (deep & wide) spacing 1000-3000μm plastic cylinder (axial)
[Neimark et al.; 2003]	sandpaper 40/80 grit size 30/60μm rms amplitude cylinder, radius 16mm (radial)	50 pin ribbon cable 0.78 features/mm cylinder, radius 16mm (radial)
[Arabzadeh et al.; 2005]	compact disk	sandpaper P100, P280, P400, P1200 mean grain diameter 162μm, 52.2μm, 35.0μm, 15.3μm
[Albarracín et al.; 2006]	flat metal surface Roughness $Ra = 0.11\,\mu$m flat acrylic surface Roughness $Ra = 0.3\,\mu$m	flat wood surface Roughness $Ra = 2.52\,\mu$m Sandpaper P1000, flat carrier Roughness $Ra = 0.3\,\mu$m
[Ritt et al.; 2008]	fine milled surface flat plastic sheet	grid of squared features edge length 3000μm height max. 2000μm flat plastic sheet
[Wolfe et al.; 2008]	sandpaper P1200, P800 mean grain diameter 15.3μm, 21.8μm flat aluminum sheet	sandpaper P150, P400 mean grain diameter 100μm, 35.0μm flat aluminum sheet
[Zuo et al.; 2011]	smooth surface flat plate	grooved surface spacing 1mm, 2mm, 4mm, 8mm depth 1mm flat plate
[Oladazimi et al.; 2018]	sandpaper P1200, P600 semicircular carrier	sandpaper P80, P240, P400 semicircular carrier

The scanning velocity seems to have a major influence on the procedure. In contrast, changing the distance to the object changes the discriminability in a way that a successful detection is always possible, but it depends on the used method to extract the stick-slip events. Similar findings are reported in [Albarracín et al.; 2006]. The root mean square levels of the recorded signals were used to determine the roughness of the surface, whereby a closer distance to the object has a positive influence on the discriminability. Furthermore, a larger root mean square level corresponds to a larger roughness of the object, see Table 2.1. Extending this approach, also sandpapers of different grain sizes were classified but in this case, a large distance to the object had a positive influence while discriminating very rough objects and a closer one is positive in the case of smooth objects [Farfán et al.; 2011]. Slightly controversial to the mentioned facts regarding the influence of the scanning speed and the object distance are the findings in [Boubenec et al.; 2014]. In this work, the surface texture is represented by irregularly spaced grooves. Contacting this surface texture with a macro vibrissa hair shaft, the oscillation amplitude is strongly influenced by the natural frequency of the vibrissae hair shaft and the texture induced part. The authors assume that a signal modulation takes place. The natural frequency carries the surface induced signal in form of a slowly (in comparison to the natural frequency) changing amplitude of the combined signal. The authors suggest that these findings are comparable to findings regarding stick-slip.
Until here, all studies were done for macro vibrissae. But, also the use of micro vibrissae in context to surface texture detection is reported in literature [Kuruppath et al.; 2014].

Summarizing, like in the previous discussion of, e.g., object localization, animals seem to use complex scanning strategies that rely on different underlying mechanisms like the stick-slip effect. They can use one or multiple vibrissae and can detect very different types of surface textures. Until today, it is not totally known how animals evaluate a surface texture.

2.1.4 Functional architecture

Nature optimizes every part and component of living things every day. So, the particular structure and properties of vibrissae are developed for a special reason. The following subsection tries to identify the effects of the properties mentioned in Subsection 2.1.1 considering the findings of Subsections 2.1.2 and 2.1.3.

The vibrissa hair shaft gathers the physical stimuli during tactile exploration, so its shape seems to be optimized to fulfill this task [Maravall, Diamond; 2015]. But it does not only gather the stimuli, it transmits the stimuli to the FSC and assists to convert the physical stimuli into mechanical energy [Diamond, Arabzadeh; 2013], [Bagdasarian et al.; 2013], [Sofroniew et al.; 2015].

If a vibrissa hair shaft gets deflected by touching an object, the spatial location of the contact gets transformed into mechanical energy (potential energy) due to the corresponding deflection of the vibrissa hair shaft. The deflection of the vibrissa hair shaft can be actively controlled by the chosen whisking pattern, see Subsection 2.1.2. Getting closer to an object, a vibrissa hair shaft gets more deflected, and consequently, there is more potential energy. The relation between the magnitude of deflection and resulting energy changes during a sweep dependent on the trajectory of the vibrissa hair shaft and object shape [Hipp et al.; 2006]. The magnitude of deflection seems to be optimized [Carvell, Simons; 1995], [Mitchinson et al.; 2007] to maximize the quality of performance of particular tasks, see Subsection 2.1.3. For example, for large deflections of a vibrissa hair shaft, it is reported that adjacent vibrissae get affected due to the deformation of the surrounding skin [Bosman et al.; 2011]. So, the shape of each vibrissa hair shaft contributes, as well as supports the signal processing and its overall slenderness and elasticity are necessary to sense even very small surface textures [Carvell, Simons; 2017]. In context to sensing surface textures, the authors of [Hipp et al.; 2006] mentioned that in principle to sense a surface texture no special shape of a vibrissa hair shaft is necessary, but that the typical shape of a vibrissa hair shaft causes a band-pass like filtering of the gathered signals. Already in [Neimark et al.; 2003], the band-pass like behavior of a vibrissa hair shaft is reported and, furthermore, a signal amplification due to the vibrissa hair shaft and the corresponding natural frequency is supposed. This finding is controversially discussed because an amplification of the captured signal, while the excitation frequency is close to the natural frequency of the vibrissa hair shaft, is inevitable and therefore maybe not an underlying mechanism to shape the captured signal [Arabzadeh et al.; 2009]. The overall slenderness of a vibrissa hair shaft is affected by the change of its diameter (taper) from base to tip, see Subsection 2.1.1. In addition as well as in contrast to the view of the authors of [Hipp et al.; 2006], there are various advantages of the tapered shape beyond a simple band-pass characteristic. Firstly, the findings of [Solomon, Hartmann; 2011] and [Huet et al.; 2017] regarding the influence of a tapered shape must be emphasized. Analyzing the relation between the location of a contact point and the corresponding support reactions in 2D- or 3D- space [Solomon, Hartmann; 2011], [Huet et al.; 2017], it is reported that in the case of a tapered shape only two (2D-space) or three (3D- space) support reactions have to be evaluated in order to determine the location of the contact point. These findings are restricted strongly by the used theoretical model and assumptions, e.g., neglecting friction in the contact point. Consequently, at a second view, these findings must be seen in a crucial perspective even if other works discuss a major influence of the tapered shape for object locating tasks [Williams, Kramer; 2010], [Pammer et al.; 2013], [Behn et al.; 2017a]. Comparing a tapered vibrissa hair shaft with a hypothetical, cylindrical one while scanning an object, the tapered vibrissa hair shaft detaches from the object earlier than the cylindrical one for an equal induced displacement [Quist, Hartmann; 2012]. But the slip-off of the vibrissa hair shaft from the object is smoother if the vibrissa hair shaft is tapered [Hires et al.; 2013].

This effect is caused by the reduced bending stiffness near the tip of the hair shaft compared to its base. Furthermore, the reduced bending stiffness results in smaller contact forces and a strong deflection of the tip section whereby the base section remains nearly undeformed [Williams, Kramer; 2010], [Sofroniew et al.; 2015]. The sensitivity and geometrical dimension of the tip section are related to enhanced performance in sensing fine surface textures [Williams, Kramer; 2010]. Furthermore, the geometric dimension of the tip section prevents the vibrissa hair shaft to get clamped [Sofroniew et al.; 2015] between grooves and gaps. Also, an influence on the behavior of the hair shaft exposed to the stick-slip while scanning a surface texture is shown. Here, the tapered shape protects the vibrissa hair shaft to get stuck on the surface, it enables a smooth displacement across the surface texture [Hires et al.; 2013]. Another advantage of a tapered shape is that the natural frequency of a vibrissa hair shaft is robust against damages of the tip [Williams, Kramer; 2010].

Another prominent feature of a vibrissa hair shaft is its inherent curvature, see Subsection 2.1.1. During tactile exploration, a rat can orientate a vibrissa hair shaft in two ways, pointing the concave side of the inherent curvature in displacement direction - concave forward; or pointing against the displacement direction - concave backward.

It is worthy to mention that these two modes are not necessarily related like: protraction - concave forward; retraction - concave backward. The muscles connected to the FSC allows for both: a protraction with concave forward as well as backward orientation of the inherent curvature. Most works do not mention exactly how the curvature of the vibrissa hair shaft was oriented. But, in [Carvell, Simons; 1990], [Grant et al.; 2008] protraction is related to a faster velocity than retraction. The authors of [Carvell, Simons; 1990] interpret this difference as a result of the different muscles used for pro- and retraction. However, in [Grant et al.; 2008], it is mentioned that this difference is purposed because it may enables fine discrimination of the surface texture, see Subsection 2.1.3.

Nevertheless, the orientation of the curvature has a strong influence on the captured signal and behavior of the vibrissa hair shaft dependent on the shape of the scanned object. Touching an object with a vibrissa hair shaft in concave forward orientation leads to longer contact phases until the vibrissa hair shaft snaps-off the object than in concave backward orientation [Quist, Hartmann; 2012]. But, also when the vibrissa hair shaft is swept concave backward along the object, the scanned part of the object shape increases compared to a hypothetical, cylindrical vibrissa hair shaft [Behn et al.; 2017b]. The orientation of the inherent curvature influences the values of the captured signals/support reactions caused by an object contact as well. The bending moments are less affected than the transversal and axial forces. In case of the concave forward orientation, the transversal forces are more affected than the axial ones. In contrast, for the concave backward case, the axial forces are more influenced.

Generally, the concave forward orientation leads to larger support reactions than concave backward orientation because the vibrissa hair shaft gets straightened first and then further bent [Hires et al.; 2013]. The orientation of the curvature is important for texture scanning as well. Because a concave backward orientation prevents the tip to push into grooves or gaps and getting clamped. Summarizing, the findings regarding the orientation of the inherent curvature, it must be highlighted that there is an anisotropic behavior of the captured signals in dependence on the orientation and that orientating the vibrissa hair shaft in concave forward direction leads to a signal amplification. Moreover, the influence of the inherent curvature on the determination of a location of a contact point in 3D- space was discussed by the author of the present work and his colleague Lukas Merker from Technische Universität Ilmenau. For a straight vibrissa hair shaft, it is demonstrated that the location of a contact point can be determined by reducing a 3D- mechanical model to a 2D- one [Clements, Rahn; 2005], [Merker et al.; 2019]. This will be impossible using an inherently curved vibrissa hair shaft because of a torsional moment around the longitudinal axis of the vibrissa hair shaft. To overcome this issue, the plane of the inherent curvature must be equal to the deformation plane in consequence of a load/contact. Since rats can rotate their vibrissae it seems to be possible that they control this torsional movement in a way that the torsional moment gets minimized until the curvature and deformation plane are equal. This idea falls in the fact that a part of the mechanoreceptors shows a strong angular tuning [Furuta et al.; 2020]. Angular tuning means that only mechanoreceptors located in the direction of the stimuli respond to the stimuli. So, a 3D-location of a contact can be determined by evaluating the axial force, the transversal force, the bending moment, and rotation angle set by the muscles. In contrast, neglecting the rotation angle, at least two components of the transversal force must be measured and evaluated which seems to be unlikely considering [Furuta et al.; 2020].

Also, the layered structure of a vibrissa hair shaft is important for its deformation. The scales on the outside of a vibrissa hair shaft (see Subsection 2.1.1) are touching the object while tactile exploration. There is nearly no information about the function of scales in general neither for their influence on tactile sensing. In [Yanli et al.; 1998], the authors assume that the function of the scales is to avoid strong wear and to protect the inner parts of the vibrissa hair shaft and that the overlapping of the scales is a result of a fast growing of the vibrissa hair shaft. As far as the author knows, the only hypothesis regarding an influence of the scales on tactile scanning with vibrissae is reported in [Voges et al.; 2012]. Here, the scales are associated with both an increasing bending stiffness and a gear rack that induces high-frequency vibrations to the vibrissa hair shaft while scanning a surface. The thickness of the cuticle and cortex and the presence of the medulla are generally related to support the overall flexibility of the vibrissa hair shaft [Voges et al.; 2012], [Maravall, Diamond; 2015]. The influence of the medulla is discussed in [Yang et al.; 2019].

The authors compare a hypothetical vibrissa hair shaft without medulla to a vibrissa hair shaft with medulla constant and conclude that the medulla optimizes the relation between the second moment of area and moment of inertia. For example, the increase of the second moment of area at the base of the vibrissa hair shaft leads to a smaller deflection under deadweight. Furthermore, the decrease in weight because of the approximately hollow structure of the medulla causes a lower energy to move a vibrissa hair shaft. Even if this effect is small, it sums up over multiple vibrissae and whisks. Not related to any of the layers of a vibrissa hair shaft but mentioned as overall property, the authors of [Hartmann et al.; 2003] found that a vibrissa hair shaft is notably damped. They assume that this strong damping is caused by viscous damping, e.g., of the surrounding air as well as by hysteretic damping of the biological material. The strong damping is necessary to ensure that a vibrissa hair shaft stays in contact with an object after initial collision and does not start banging on and off.

When it comes down to the influence of the structure of the FSC on the captured signals and overall behavior of the vibrissa, most of the available information is unverified. Across different species, different sizes of FSCs are reported [Bosman et al.; 2011]. The size of an FSC seems to change for, e.g., different environmental conditions. Thus, an FSC of seals is much larger than the one of rats. To maintain the operating temperature of an FSC in an aquatic environment, the FSC of seals shows larger blood vessels than the one of rats. This is necessary since the function of mechanoreceptors is temperature dependent [Dehnhardt et al.; 1998], [Soderquist; 2002], [Erdsack et al.; 2014]. Since also rats are exposed to a changing temperature, the presence of these blood vessels, even if they are smaller than the ones of seals, seems to be necessary to keep a vibrissa working in cold environments with a good performance.
Especially, the blood sinus is the subject of various hypotheses. Next to maintaining a constant temperature of the mechanoreceptors, the stiffness of an FSC seems to be adjustable by controlling the blood pressure, e.g., [Bosman et al.; 2011], [Carl et al.; 2012]. In doing so, the overall stiffness of an FSC can be changed [Towal et al.; 2012] and, therefore, also its influence on the scanning process because the stiffness of the support of the vibrissa hair shaft has a major influence on the deformation and related signals [Ritt et al.; 2008]. The vibrissa hair shaft and the supporting FSC are approximately fixed to each other (see Subsection 2.1.1), this is assumed to be necessary to distinguish between axial and transversal forces. This procedure may be further supported by the spatial distribution of the mechanoreceptors and the surrounding tissues which have maybe different stiffnesses. Hence, controlling the stiffness of the blood sinus will also affect the response of the mechanoreceptors [Rice et al.; 1986] in order to tune the captured signal into the operating range of the mechanoreceptors [Bosman et al.; 2011]. Generally, there seems to be a relation between type of the mechanoreceptor, its sensitivities for different mechanical stimuli, e.g., bending moment or axial force, and the surrounding tissue [Furuta et al.; 2020].

Another way to manipulate the stiffness of an FSC is to deform the complete FSC/follicle capsule by contraction of the connected muscles which maybe increase the sensitivity of the mechanoreceptors as well [Carvell, Simons; 1995]. This becomes important with view to the startle and exploration mode. While active whisking, the stiffness of the FSC probably changes, whereas in startle mode there is no active movement of the vibrissae hair shafts and, therefore, no influence of the activated muscles on the FSC [Ritt et al.; 2008], [Maravall, Diamond; 2015].

Mechanoreceptors can be active/*switched on* or adapted/*switched off* [Soderquist; 2002]. To activate them the authors of [Towal et al.; 2012] assume three mechanisms: the mechanoreceptors can be activated by a collision with an object, by squeezing from the surrounding muscles, or by a change of the volume of the blood sinus. The authors of [Estebanez et al.; 2018] further discuss if one vibrissa gets deflected the adjacent vibrissae get activated. In [Szwed et al.; 2003] it is reported that mechanoreceptors embedded in the blood sinus do less respond to free air whisking and will be activated by touching an object. In contrast, the mechanoreceptors located in the region of the rete ridge collar and the inner- and outer conical body are sensitive to signals induced by free air whisking. This idea of a functional grouping of mechanoreceptors is extended with view to the innervation of an FSC. The upper part of an FSC, including the rete ridge collar and the inner- and outer conical body is innervated by the superficial vibrissal nerve and the lower part by the deep vibrissal nerve.
In general, there are two types of mechanoreceptors (see Subsection 2.1.1), the SA and RA mechanoreceptors work like a low/high-pass filter [Hipp et al.; 2006], [Claverie et al.; 2017]. They split the signal in a low-frequency and a high-frequency part, e.g., [Mitchinson et al.; 2004]. This signal decomposition can be further observed in neuronal recordings [Stüttgen et al.; 2006]. Along the complete signal pathway, multiple times, sophisticated pre-processing takes place, until finally, the information reaches the central nervous system [Bagdasarian et al.; 2013]. For example, the connected neurons are filters, thresholds, and pre-processors [Diamond; 2010].

Whisking is actively controlled as described in Subsection 2.1.2, furthermore, the sensory information is combined in a closed-feedback-loop with the motor control, e.g., [Ahissar, Oram; 2015].
Therefore, it is assumed that the whisking signal and the object contact induced signal are in anyway compared to identify the information related to the object [Diamond; 2010], [Ahissar, Knutsen; 2011]. This idea is negated in [Garrett; 2009], the author discusses that the information related to the object do not change for different ways how to touch it and, consequently, there is no need to know anything about the motor output. But, a connection between motor control and sensor allows for an integration of the captured signal across various whisks [Sofroniew et al.; 2015].

Finally, also the complete array of vibrissae exhibits functional properties [Hobbs et al.; 2016]. The different types straddlers, macro, and micro vibrissae are used in different situations, e.g., [Brecht et al.; 1997], but not necessarily to fulfill different tasks, see Subsection 2.1.3. Also, inside the group of macro vibrissae there are, maybe, functional groupings, the authors of [Hobbs et al.; 2016] reported functional differences between rostral macro vibrissae and caudal ones for spatial distance determination. Generally, there is evidence that if more vibrissae are in contact with the object the sensing task is performed more accurate, e.g., [Carvell, Simons; 1990].

2.2 Classification and adaption

The findings of the previous Section 2.1 are taken together in order to combine and classify them into the scheme of a biomechatronic system like introduced in [Witte, Schilling; 2017], see Subsection 2.2.1. Afterwards, the in- and output signals of the biomechatronic system are analyzed in detail in Subsection 2.2.2.

2.2.1 Biomechatronic system

The top part of Figure 2.7 is adapted and only slightly modified from [Witte, Schilling; 2017]. The biomechatronic system consists of two major parts. One part is the global control. In biology, the global control is represented by the central nervous system. As mentioned earlier, the neurobiological periphery of the vibrissae, including the central nervous system of the animal, is not analyzed and considered in the present work. Therefore, it is also not mirrored in detail in the biomechatronic system. The global control is connected by nerve fibers (environmental information interface) to the local control. In the case of vibrissae, these nerve fibers are efferent (actor) and afferent (sensory) nerves. The connected local control cannot be clearly related to any component of the vibrissal system. The authors of [Kleinfeld et al.; 2006] show that various control loops are passed through before the information reaches the cortex. For example, the so-called brainstem loop includes and connects the vibrissae, the trigeminal ganglion, the trigeminal nuclei, and the facial nuclei/reticular nuclei. So, there is a control loop between the sensor and actor pathway before the central nervous system. The local control sends information to the actor as well as it receives information from the actor. In the natural system, the actor can be represented by different components. The vibrissae can be moved by contraction of the muscles connected to the vibrissae (whisking, illustrated by the small orange box in Figure 2.7) or they are displaced by locomotion of the complete animal or only a head motion (illustrated by the larger orange box in Figure 2.7). In all three cases, the transmission goes through the FSC and the vibrissa hair shaft whereby a vibrissa hair shaft is an effector too. A difference between the three mentioned actuations is the sensor feedback (interoception). As described in Subsection 2.1.2, the mechanoreceptors inside the FSC capture a signal induced by whisking.

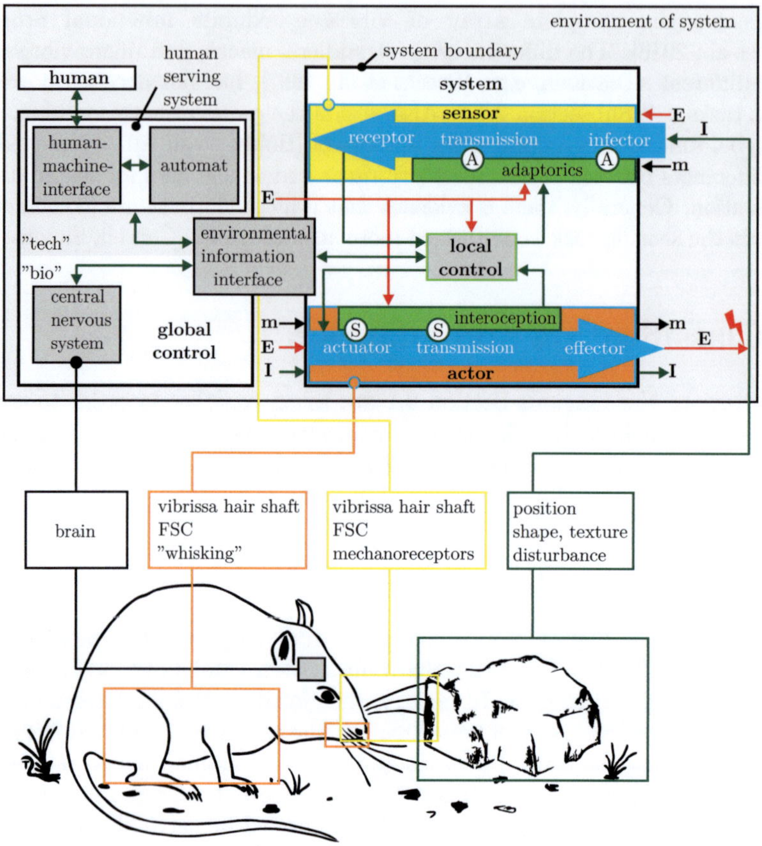

Figure 2.7 – Biomechatronic system following [Witte, Schilling; 2017]: A- actor; S- sensor; E- energy; I- information; m- mass flow.

So, it can be expected that during whisking the position or the position related signal of a vibrissa is known and can serve as a reference signal or for a signal modulation liked discussed in, e.g., [Diamond, Arabzadeh; 2013]. In the given illustration of a biomechatronic system, the actor and the sensor (exteroception) parts are clearly split. In contrast, in the biological example, both parts are strongly merged. The vibrissa hair shaft is the effector and the infector at the same time, as well as the combination of FSC and vibrissa, serves in both parts for the signal transmission. The origin of the captured information is a contact between a vibrissa hair shaft and an object. So, the vibrissa hair shaft gathers the information and transmits it to the mechanoreceptors (receptor). Since the FSC supports the vibrissa hair shaft and includes the mechanoreceptors, it has a strong influence on the transmission and is consequently part of the sensor section. The sensor section is adaptive (adaptorics) in order to increase the sensor performance.

For example, the captured signal can be tuned by controlling the displacement and orientation of the inherent curvature of a vibrissa hair shaft (adaption of the infector), see Subsection 2.1.4. The transmission gets adaptive due to the properties of the FSC, see Subsection 2.1.4. First, the transmission characteristics of the vibrissa hair shaft are dependent on the stiffness of the support/FSC. Therefore, a squeezing of the FSC by the muscles may lead to a larger support stiffness, and consequently a difference in the captured signal. Furthermore, the blood vessels inside the FSC can tune the sensitivity of the embedded mechanoreceptors as discussed in Subsection 2.1.4. The unity of sensor and actor (system) confirms the concept of a stimulus leading apparatus [Carl et al.; 2008] that allows for a closer identification of the functions of the different components, see Figure 4.1.

The local control, the sensor, and the actor parts are supplied by an energy flow/input (E), e.g., the release of nutrients or oxygen to maintain them working. For example, oxygen is carried by blood flow (mass flow).

The energy input has a second function, it also can provide information. For the actor, the energy signal contains information, e.g., which muscles should be contracted in order to fulfill a desired displacement of a vibrissa hair shaft. As shown in Figure 2.7, the in- and output flows of the actor are equal. In reference to a vibrissa, the mass flow (output) is absent, but the energy and information flows are present. The energy output is related to the contact to an object, and the information flow includes, e.g., the current position or displacement velocity which are observable from the outside of the system. Following the sequence of the actor part, the energy flow gets transformed. It enters the muscles as electrical energy and converts to potential and kinetic energy of the vibrissa.

The same three flows enter the sensor part of the biomechatronic system. As well as for the actor part, the mass flow (input) is absent in the case of a vibrissa. The entering information is carried by the energy flow. When the vibrissa hair shaft touches an object, the vibrissa hair shaft deforms. For example, contacting a smooth, close object can yield a large deformation of a vibrissa hair shaft. Therefore, the spatial information of the contact point location is converted: into a contact force, and consequently, into an energy of elastic deformation of the vibrissa hair shaft. This elastic energy gets further transformed, first, back to a contact force (contact between the hair shaft and FSC/mechanoreceptor), and finally, to an electrical energy, action potential respectively, see Figure 4.4. Alike observations are reported in [Bagdasarian et al.; 2013].

As indicated by a red flash in Figure 2.7, the energy output of the actor influences the energy input of the sensor. Furthermore, with regard to Subsection 2.1.3, the information that enters the sensor seems to be very variable. Therefore, the kind of information which possibly enters the sensor must be analyzed. Without a deeper understanding of the kind of the gathered information, a cognition of the underlying mechanisms supporting the sensing procedure is not possible.

2.2.2 System input

In context of tactile sensing, the system input is provided by an object and its properties whereby the properties are the searched information. So, it must be clarified which kind of properties exhibits an object and which ones are important in the present context. For example, the conductivity of the scanned object is of minor importance than the object shape. Proceeding from this, an object must be seen from three points of view: the technical description of object properties, how object properties contribute to the interaction between two touching bodies, and how does an animal feels an object. Scanning an object of interest, for each mentioned case, the scanned object is the same but the captured information can be very different.

Surface metrology:
In literature, the terms shape, macroscopic and microscopic can be found frequently in context of surface metrology. These terms refer to the size of anything regarding the object of interest. The shape of an object is its overall form like spherical, cylindrical or straight. Macro and micro originate from the Greek language whereby macro stands for adjectives like large, wide, long, and micro for small, narrow. The standard [DIN 4760] provides a scheme to classify objects based on their geometric/surface properties, see Table 2.2.

The largest geometric dimension of an object is its shape/form [VDI/VDE 2631]. The shape is affected by form errors, see Table 2.2. According to [VDI/VDE 2631], a shape can be measured by touching it with a tactile probe. A sequence of contact point locations results out of a displacement of the probe or the object. The sequence of points results in a discrete representation of the shape. In doing so, frequently low-pass filters are applied to the measured signal in order to get rid of high-frequency signal components, e.g., vibrations induced by the motor or smaller surface structures. Finally, the low-frequency part of the signal represents the shape. The smaller surface structures are described in [ISO 8785], [DIN 4760]. A surface texture is defined as a repeatable or random variance of the 3D- surface topography. The variance includes the roughness, waviness, surface imperfections, and form errors. The surface imperfections are defined in [ISO 8785]. There are 31 types of surface imperfections, like: erosion, corrosion, scales, cracks, holes, bumps among others. Beside these imperfections, components of a surface texture get classified in [DIN 4760]. There are form deviations of first to sixth order, see Table 2.2. Surface texture components of second order are an order of magnitude larger than those of third and fourth order. The difference between third and fourth order is the uniformity of the presence of the structure and not its geometrical size. In [Davim; 2010], the form deviations of order one and two are related to macroscopic deviations and the others to microscopic deviations. The form deviations of fifth and sixth order are out of scope for the present work. Each deviation can be expressed by different numerical parameters. In [Volk; 2013] approximately 38 different values are mentioned, which are commonly used in the German industry.

Table 2.2 – Overview of surface texture deviations of first to sixth order adapted from [DIN 4760].

Form deviation	Example	
1. Order: form errors		straightness, flatness, roundness ...
2. Order: waviness		waves[3]
3. Order: roughness		grooves[3]
4. Order: roughness		scales, peaks[3]
5. Order: roughness	–	change of the microstructure, e.g., corrosion ...
6. Order:	–	errors of the crystalline structure

It has to be pointed out that there is no clear definition of which texture deviation is well described by which one of these parameters. This decision is driven by the characteristics of each case and application.

A surface texture can be evaluated using different methods and devices. One example is the profile method, which is described in the following, see [ISO 4287]. Here, a stiff probe with a specified tip (see Figure 1.2) is pulled over the surface. The captured signal represents a 2D- section of the scanned surface texture including spatial frequencies and amplitudes of all surface texture components. The dimension of the tip influences the measured signal. It acts like a morphological filter. Consequently, there is already a difference between the real surface and the measured one. The measured signal is the total profile, it gets decomposed in a primary profile by filtering high-frequency components defining a cut-off wave length [ISO 3274],[ISO 4287]. The primary profile then gets further decomposed by filtering to the residual profiles like the roughness profile or the waviness profile. The applied filters are characterized by the chosen cut-off wavelength again. The chosen cut-off wavelengths and filter types have a strong influence on the results. For example, there are filters causing run-in and run-out effects that must be considered during evaluating the data. In [Krystek; 2009] the advantages and differences between various approaches of filtering the measured signal are discussed. Finally, the filtered profiles are used to calculate a numerical parameter as a representation for special surface characteristics. For example, the arithmetical mean can be calculated for the primary profile (P_a), the waviness profile (W_a), or the roughness profile (R_a). To calculate this parameter the evaluated profile gets sectioned, and for each section, the parameter is calculated. The arithmetical average of these parameters is the final result.

[3] see [ISO 8785]

It must be highlighted at this point, that talking about the roughness of a surface does not necessarily refer to the roughness profile and not even to any specific parameter. Summarizing, there is no numerical parameter called roughness and the transitions from shape to waviness and waviness to roughness are dependent on more or less freely chosen filter coefficients and cut-off wavelengths.

Tribology and contact mechanics:

Tribology and contact mechanics are likely and associated with the contact between two bodies and/or media relative moving to each other. The tribology includes friction, wear, lubrication, and interfacial interactions [DIN 50323]. For the present work, considering the findings of Subsection 2.1.3, frictional contact mechanics are most important. Topics like wear, capillary contact phenomena, Hertzian contact effects, or lubrication contribute to the phenomena observed in Subsection 2.1.3 but seem to be of minor importance for this work. Furthermore, adhesive forces are excluded as well, even if adhesive forces are often present in contacts including a biological material as one contact material [Popov; 2010]. Also, models describing the contact between two rough surfaces are neglected in a first step because the scope of these models is different from the effects discussed in Subsection 2.1.3. Therefore, the present discussion follows essentially [Popov; 2010] if not otherwise stated, and is limited to frictional contacts, more specific to dry friction being aware that there are other extended models, e.g., the LuGre friction model [Johanastrom, Canudas-de-Wit; 2008] or the Magic Formula [Pacejka, Bakker; 1992].

Dry friction describes the contact between two contacting surfaces. The contact is represented by a contact force F. Inducing a relative motion between the surfaces, F consists of a component in normal direction F_n and one in tangential direction F_t with respect to the contact plane, see Figure 2.8(a). Three main laws/assumptions introduced by Guillaume Amontons and Charles Augustin Coulomb must be considered discussing dry friction. The first of Amontons' laws states that F_n and F_t are proportional to each other, and the second one that the contact area between the touching surfaces does not influence F_t. There is no deformation or penetration of the contacting surfaces. Consequently, a surface to surface contact gets reduced to a contact in one single point, loaded by a contact force. Thirdly, Coulomb found that F_t seems to be independent of the amount of displacement velocity between the surfaces whereby F_t acts opposed to the displacement velocity.

For example, a body is connected to a spring and loaded by its weight force in normal direction while it stays in contact with a moving surface, see Figure 2.8(a). First, depending on the initial state, it is assumed that there is no relative displacement between the body and the surface. They stick to each other until a maximum of the quotient of F_t to F_n is reached. Passing this limit, the body starts moving, Figure 2.8(b).

The sticking period is related to static friction and the moving one to dynamic friction. In both cases, it is assumed that the F_t is proportional F_n as found by Amontons, see (2.1):

$$\mu = \frac{|F_t|}{|F_n|}; \qquad \mu \in [\mu_d, \mu_s] \tag{2.1}$$

The quotient of F_t and F_n is the friction coefficient μ. The upper limit of μ is the static friction coefficient μ_s and the lower limit the dynamic one μ_d. According to the classic cone of friction (Figure 2.8(a)), the theoretical maximum is $\mu = 1$ corresponds to a friction angle of $\zeta = 45°$, see (2.2):

$$\mu = \tan(\zeta) \tag{2.2}$$

For special material combinations or including adhesive forces, μ can also be larger than one. In (2.1) the influence of the direction of the displacement velocity v is missing. Therefore, it can be rewritten in the form:

$$\vec{F}_t = -\mu|\vec{F}_n|\frac{\vec{v}}{|\vec{v}|} \tag{2.3}$$

to express F_t as function of μ, F_n and v. The transition from sticking to sliding is characterized by an abrupt change of F_t, followed by a constant trend corresponding to F_t with respect to μ_d. In the case that the external force is changing, it is possible that the body gets stuck again after a sliding period, see Figure 2.8(c). Here, first F_t reaches a maximum for the first transition from sticking to sliding and the abrupt change takes place. Recurring sticking and sliding cause a periodic signal in dependence on a change of the external force. This behavior is called stick-slip (sometimes also called sprag-slip). Generally, a changing load is required to generate stick-slip displacement. The load can change in tangential or normal direction.

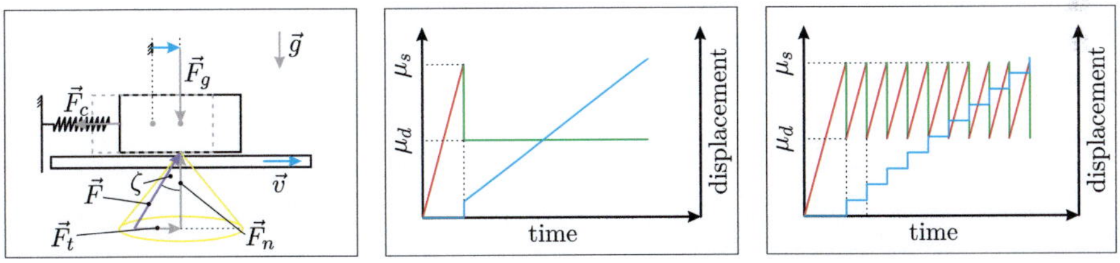

(a) **(b)** **(c)**

Figure 2.8 – A typical example for a tribological system is the friction oscillator in **(a)** with $\vec{F}_c$; spring force; $\vec{F}_g$ as weight force; $\vec{g}$ as gravity constant; $\vec{v}$ as velocity; ζ as friction angle; $\vec{F}_t$ as tangential force; $\vec{F}_n$ as normal force and $\vec{F}$ as contact force. The yellow cone represents the friction cone. In **(b)** and **(c)** typical trends for the friction coefficient μ are illustrated. The transition from sticking (red) to sliding (green) indicates a stick-slip motion (cyan). The figures are qualitative and adapted from [Enger, Feinle; 2014].

Hence, the influences of the surface properties and further conditions on μ must be analyzed because μ is directly related to the contact force. In contrast to Coulomb's Law, it is found that F_t is dependent on the amount of the displacement velocity. For velocities below a critical value, an unstable displacement of the system is possible. Similar observations are done for the influence of the stiffness of the system. A stiffness below a critical value causes an unstable displacement. Regarding the surfaces and their properties, a longer duration of sticking can correspond to larger μ_s. Furthermore, for elastic bodies, the linear relation between F_n and F_t becomes inaccurate because the assumption that the contact can be reduced to a contact in one point is not valid anymore, since the real contact areas can strongly increase. In mechanics, a contact between two bodies considering friction is often classified as rough and one with negligible friction as smooth. This classification implies that the surface roughness determines the magnitude of the present μ. In general, there is no direct relation between the surface roughness and μ. For example, a combination of a very smooth surface can exhibit a larger μ than a combination of rough surfaces. In conclusion, a μ corresponds to a specific combination of contact partners considering specific conditions.

Following in Subsection 2.1.3, vibrissa regarding findings demonstrate that the properties of an object can be evaluated by recording the support reactions during touching an object with a vibrissa. At least the idea of determining the information about the surface texture or roughness must be seen as very critical considering the discussion above. The author of [Wiertlewski; 2013] confirms this statement in context to the tactile perception of finger tips.

Objects and surfaces in context of vibrissae:
The relation between provided information/object property and gathered stimulus is crucial in the analysis of tactile perception with vibrissae. Actually, this point is widely unknown. The discussion in Subsection 2.1.3 indicates the problem. For example, observing a vibrissa touching an object by recording the neuronal response of the mechanoreceptors can indicate a contact. But if this response is related to the shape of the object or its texture is difficult to determine. Varying the texture, e.g., by changing the grit size of a touched sandpaper the effect of different grit sizes on the captured signals can be evaluated, e.g., [Wolfe et al.; 2008]. Here, a first problem arises across available literature regarding texture perception with vibrissae. Like exemplarily summarized in Table 2.1, the experimental setups of different works analyzing texture perception are extremely different. It seems to be questionable if the results of scanning an u-shaped object covered with sandpaper are comparable to those generated by using a flat carrier covered with sandpaper because the information of the object shape interferes with the information of the surface texture. A first attempt in this direction is done in [Hipp et al.; 2006]. The authors report that there is a low-frequency signal part that belongs to the shape of the object and a high-frequency part belonging to its texture. Due to the low/high-pass characteristics of the mechanoreceptors inside the FSC, the signal components may get decomposed, but this is hypothetical. Furthermore, also reviewing only the used surfaces to analyze the process texture detection with vibrissae, huge differences can be found. First, the used terminology is not consistent between different works and causes confusion by comparing them. The work [Isett et al.; 2018] distinguishes between local surface features and macroscopic texture. Here, local surface features have a geometric dimension in the magnitude of millimeters, and the macroscopic surface texture is made of different sandpaper, e.g., P150 (mean grain diameter 100 μm). In contrast, in [Schwarz; 2016] the surface roughness is mentioned to be microscopic and macroscopic refers to the object shape and location. This comparison reveals another point, the term roughness is frequently used as a representation for a rough surface but only in a qualitative manner or it is merged with the term texture, e.g., [Jadhav, Feldman; 2010]. There are also works associating sticking of the vibrissa hair shaft in grooves and gaps to the stick-slip effect, e.g., [Zuo et al.; 2011], which ignore that the stick-slip effect is caused by friction and not by hampering of the displacement by an obstacle.

Besides these issues, texture perception is often related to the stick-slip effect as mentioned in Subsection 2.1.3. Taking into account the diverse experimental setups, this effect was observed under very different conditions and it can be assumed that it has an important influence on the process [Diamond et al.; 2008a]. But as discussed early in the present section, the stick-slip effect depends on friction and friction is not necessarily related to any surface texture parameter. So, a supposed relation between the grit size of a sandpaper and the stick-slip effect is weak. There is no information about the friction coefficient for a vibrissa hair shaft-sandpaper-contact pair.

For human hair friction coefficients ranging from approx. 0.1-0.3 are reported for different contact pairs [Bhushan et al.; 2005]. Furthermore, it is reported that there is a directional friction effect. The friction coefficient is larger moving the hair in *base to tip* direction than in *tip to base* direction [Robbins; 2012]. This effect may apply also for vibrissae even if the scales of a vibrissa hair shaft tend to be smaller compared to body hairs [Voges et al.; 2012], compare Figure 2.3(b). The author of [Popov; 2010] mentions that for biological systems additional adhesive forces are normally present.

Summarizing, it is not exactly clear which effects or information represent an object for an animal, neither which mechanisms drive object and surface texture detection. But the reviewed literature shows that there are three properties of an object that must be distinguished: the shape; surface features smaller than the overall shape but still large; and very small surface structures. The very small structures seem to contribute to the friction coefficient between the vibrissa hair shaft and object surface, and the frictional contact transmits the information about the surface texture to the vibrissa hair shaft, finally to the mechanoreceptors. But, the stick-slip is not necessarily a representation of the small surface texture features, it can be influenced by various effects.

3 Objective

A tactile characterization of unknown objects or samples is a frequently used procedure in robotics, surface metrology, production processes, daily life, and nature as well. Tactile sensors are generally involved in three types of tasks: response, manipulation, and exploration [Cutkosky et al.; 2008]. Following the authors of [Cutkosky et al.; 2008], response refers to the interaction with an external agent, manipulation means that a tactile sensor is used to determine, e.g., the necessary contact force to grasp an object, and exploration is correlated to the detection of information about the object. In engineering, commonly used sensor devices are optimized to detect specific information. For example, in surface metrology, a profilometer is typically used to determine parameters of the surface texture like the roughness. But, it is not possible to determine the present friction coefficient at the same time, for this case normally a tribometer is needed. An alike example is the somatosensory system of, e.g., rats, including the mystacial vibrissae. Each of the tasks, mentioned by [Cutkosky et al.; 2008], can be found in context to natural vibrissae. Vibrissae are part of the social interaction of animals [Ahl; 1986], which is alike to the interaction with an external agent. Current findings in literature report that rats use these tactile hairs to determine the distance to an object (manipulation), to recognize the object shape, or detect elements of the surface texture right up to microscopic dimensions (exploration). The hair shaft misses any sensory innervation, it transmits only the recorded information from the contact point to the hair follicle where finally, mechanoreceptors (sensory innervation) gather the information. In contrast to the example out of engineering, the natural example combines all the mentioned tasks in one system and executes them with high precision. The combination of functionalities and performance quality must lead to the conclusion that natural vibrissae are a sophisticated blueprint for a novel concept of an artificial tactile sensor.

Hence, the present work is a further step on the way to the overall goal to design an artificial, tactile sensor inspired by a natural vibrissa.

The structure of this work is summarized in Figure 3.1. In Chapter 4, the natural example is analyzed in detail based on findings in literature to identify important structural properties and procedures of the scanning processes. Analyzing these findings, the most important properties and procedures are determined in order to reduce the complexity of the natural example. This is necessary because the somatosensory system / natural vibrissa and the behavior of the animals during scanning an object are not totally understood until today and there are technological limitations regarding a first technical implementation in experiment. Based on the resulting findings, a theoretical model of the artificial sensor and scanning process is generated. This model is used for a set of preliminary studies to, e.g., design the test objects in a suitable way.

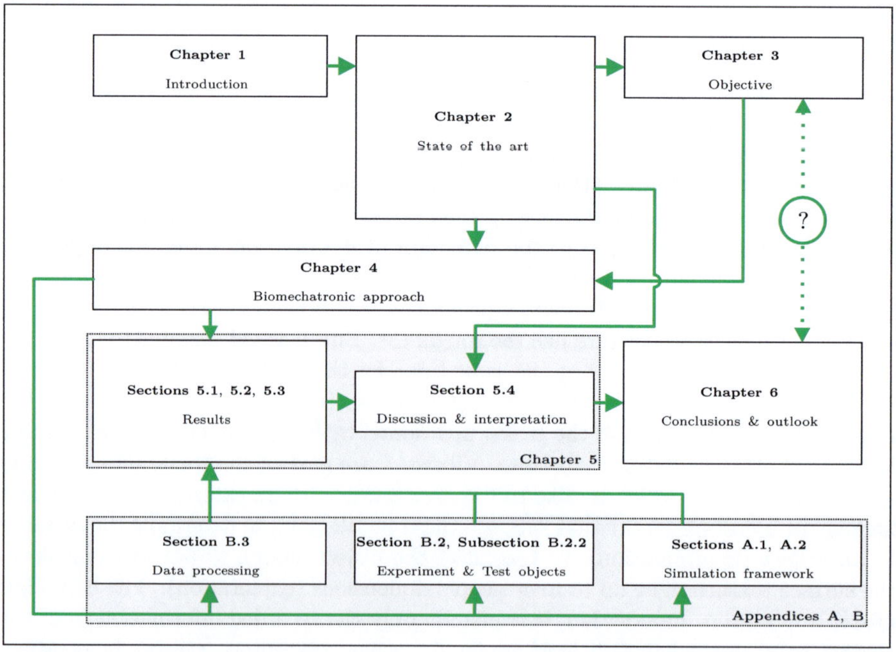

Figure 3.1 – Structure of the present work, the information flow is indicated by the green arrows.

Ensuing from these analyses, the used sensor shape, support, and structure is defined in order to set up an experiment.

In Chapter 5, objects with different contours and surface textures are scanned in simulation and experiment. Here, the decomposition of the signals recorded at the support of the sensor is in focus. All information about the object, the object contour, the surface texture features in macroscopic and microscopic dimension/friction are combined and transmitted in one signal gathered by the sensor shaft. This signal is recorded at the support of the sensor shaft and finally represented by the support reactions. They contain all the mentioned information. But, how this different information is presented in the support reactions is not clear. So, this work presents an encoding and classification scheme in order to decompose the signal into components belonging to the different information about the object. Furthermore, the inherent characteristics of the artificial sensor in combination with the applied scanning procedure are analyzed considering the goal to separate the different types of information of an object.

These steps of investigation increase the understanding of the inherent functionalities of the artificial sensor inspired by natural vibrissae and in parts of the natural example as well out of a *Reverse Bionic* perspective. Especially, the second step has to be seen and interpreted in the light of the concept of a stimulus leading apparatus.

4 Biomechatronic approach

Before the influences of different object properties on the captured support reactions can be analyzed, the inherent characteristics and the properties of the artificial sensor must be determined. What do inherent characteristics mean? It refers to effects that are influencing the captured signal, and which are caused by the sensor concept and design and not by any particular object property in order to support a decomposition of the recorded information. So, there can be a change in the signal but not a change of the touched surface. Furthermore, for the scope of the present work, the representation of an object and its surface texture must be defined in order to provide recognizable and appropriate experimental conditions.

4.1 The stimulus leading apparatus

As discussed in Chapter 2, a vibrissa is a very sophisticated sensor. Its structure fulfills multiple functions whereby there is no fixed relation which component contributes to which function. In this sense, a first classification is done in Section 2.2. There, a complete vibrissa is described using the idea of a biomechatronic system [Witte, Schilling; 2017]. Since an objective of the present work is to evaluate the influences of the mechanical structure and related mechanisms of a vibrissa-inspired sensor during tactile exploration, the functions of the structural components of a vibrissa must be clarified to design the artificial system and theoretical model in a sufficiently detailed way. Moreover, functions that might be useful for real world applications of the sensor system, e.g., a thermo-regulation [Dehnhardt et al.; 1998], [Erdsack et al.; 2014], are excluded because they are probably not in the scope of the present work. The functions of the different structural components are roughly summarized in Figure 4.1. Figure 4.1 is consistent with the idea of the stimulus leading apparatus.

The authors of [Carl et al.; 2008] propose that a natural sense organ is optimized in a way that the captured stimulus is strongly pre-processed before it finally reaches the receptor. In the present case, this means that all structural components of a vibrissa that are incorporated in the signal acquisition are part of the stimulus leading apparatus. Furthermore, also behavioral aspects like whisking must be included in this discussion. An analysis of the function-component relation shows which components must be considered for the theoretical model and artificial design of the sensor. Figure 4.1 illustrates the vibrissa components in its roles as infector, effector and for transmission (compare Figure 2.7). The vibrissa hair shaft is used to gather the stimuli by generating contact with the object of interest due to an active displacement of the vibrissa hair shaft (whisking). Therefore, a coarse positioning of the vibrissa hair shaft is necessary in order to ensure that the vibrissa hair shaft touches the object during whisking.

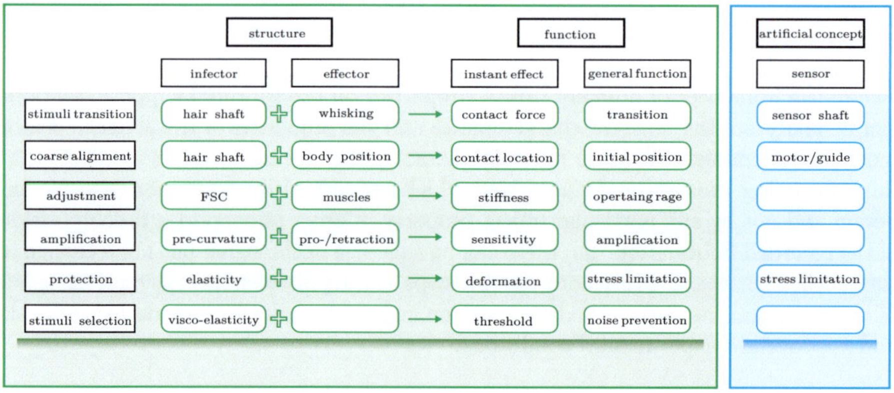

Figure 4.1 – The scheme of the stimulus leading apparatus highlights the functions of the incorporated structural components. A transfer from the natural to the artificial system, from left to right: structural components of a vibrissa; functions and influences regarding the signal acquisition and mirrored, reduced, artificial design. The figure is adapted from [Carl et al.; 2008].

This initial position can be reached by an overall body or head displacement of the animal. The architecture of the FSC may allow for an adjustment of the sensitivity of the vibrissa. As discussed in Subsection 2.1.4, it is assumed that the blood sinus change the stiffness of the FSC and, consequently, the captured signal gets affected since the stiffness of the FSC influences the magnitude of the contact force. This applies equally for a change of the stiffness of the FSC due to contraction of the surrounding muscles. The findings of [Quist, Hartmann; 2012] demonstrate that the orientation of the inherent curvature of the vibrissa hair shaft can be used to discriminate finer surface textures when the concave side is moved forward than for a backward displacement.

This change in discriminability can be seen as amplification of the signal because the contact force changes from one to the other displacement direction. In all mentioned cases, the elasticity of the vibrissa is of high importance. In addition, the elasticity is a protection for the sense organ because it prevents the vibrissa hair shaft to be broken or getting pulled out of the FSC. Here, the tapered shape contributes in both aspects as well (not included in Figure 4.1). It increases the elasticity compared to a hypothetical cylindrical shaped hair shaft and prevents the vibrissa hair shaft to get stuck in surface structures and allows for a smoother snap-off of an object [Hires et al.; 2013]. Furthermore, the diameter of the tip of a vibrissa hair shaft is remarkably small, see Figure 1.2(c). So, fine surface features can be detected as discussed in [Williams, Kramer; 2010], but simultaneously, the diameter is also a threshold for the size of recognizable surface features, e.g., a groove with a smaller diameter than the tip diameter cannot be detected. Another threshold is the visco-elasticity of a vibrissa. Out of the perspective of mechanics, a vibrissa is a damped system because of the structural properties of the vibrissa hair shaft, FSC and surrounding skin. Moreover, the mechanoreceptors are arranged in a special distribution along the vibrissa hair shaft, see Figure 2.4. They are embedded in different components of the FSC and, thereby, the stiffness of the surrounding tissue differs for different mechanoreceptors. This is important because the mechanoreceptors are activated by deformation in consequence of a contact force which in turn are influenced by the stiffness of the tissue. Consequently, the surrounding tissue may support a selection of particular stimuli, e.g., a stimulus with a very small amplitude will be easier to recognize by a mechanoreceptor embedded in stiffer tissue. Concluding, the mechanoreceptors alone are do not allow for precise measurements, but in combination with the connected periphery, the complete assembly becomes a very sensitive sense organ.

4.2 The artificial sensor design

It is obvious that the vibrissa hair shaft, the FSC and the actuation of the vibrissa are the key components. Following the idea of the stimulus leading apparatus, they determine the inherent characteristics of vibrissa. The inherent characteristics depend on two groups of parameters, see Figure 4.2. First, there are fixed parameters that cannot be changed during a scanning sweep, e.g., the length of the vibrissa hair shaft. But once they must be determined in order to design the sensor shaft for the following investigations and experiments. These parameters describe the mechanical representation of the sensor shaft. The fixed parameters in combination with the variables that correspond to the actual scanning sweep, e.g., the displacement velocity of the support v causes the inherent characteristics of the complete system. Besides, there are unavoidable deviations like vibrations in the drive. The present system is not totally unknown, it is not a *Black Box*, because the sensor shaft can be described by known beam theories and the contact between the object and the beam by tribology models and contact mechanics.

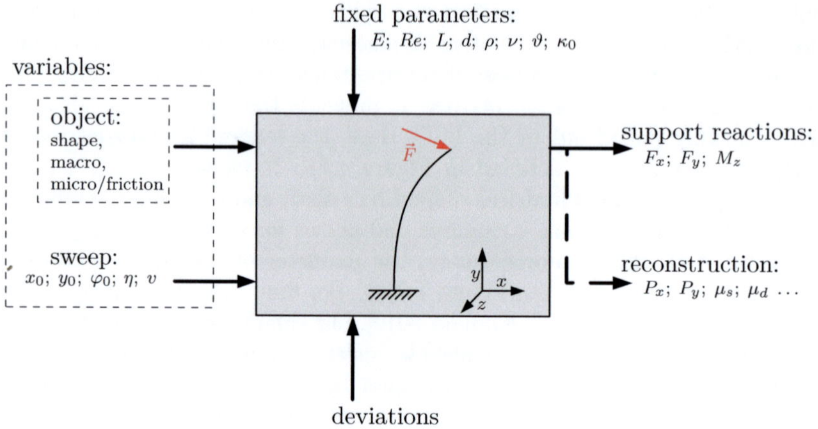

Figure 4.2 – Grey Box, deviations can include: excitations by the motor, power line frequency (50 Hz hum), acoustic emission, frequency induced by the environment, temperature changes, elastic deformations of components which are assumed to be rigid, surface texture of the steel filament since it is just assumed to be smooth.

But, these models are restricted in multiple ways so that figuratively speaking the present system is a *Grey Box*, see Figure 4.2. In this way, analyzing the inherent characteristics improves the comprehension of the system and, finally, the relation between object properties and captured support reactions can be investigated in a reliable manner.

Starting to adapt the natural paragon to an artificial sensor design, multiple simplifications and assumptions must be done:

- All interactions and displacements are restricted to take place in the x-y-plane because scanning tasks (see Subsection 2.1.3) can be reduced or transformed to 2D-tasks in most cases [Clements, Rahn; 2005]. This statement has no universal validity but applies for the examples of the present work.
- The sophisticated morphology of a vibrissa hair shaft is reduced to a straight, slender, cylindrical filament consisting of a linear-elastic, isotropic material (Hooke's material). Note, this is not even close to the structure of a natural vibrissa hair shaft. But technological limitations and considerations for the following experiments lead to this choice. For example, the layered structure of a vibrissa hair shaft is poorly understood and will result in further particular effects that may be not in the scope of the present work. Furthermore, considering a scaly surface will affect the signal because μ_s and μ_d depend on the orientation of scales, see Subsection 2.2.2. Also, this effect is not well analyzed yet and, consequently, excluded in a first step.

- The inherent curvature of a vibrissa hair shaft is already analyzed in literature and seems to amplify the signal among others. It is neglected because the sample objects for experiment are designed in a way that all object properties can be well detected and, therefore, e.g., a signal amplification is not necessary. Thus, the inherent curvature is $\kappa_0 = 0$.

- The tapered shape of the vibrissa hair shaft contributes to the protection of the vibrissa hair shaft and to the detection of small surface features. Both are not essential for the present case. The artificial sensor is used in well protected laboratory conditions and, as mentioned in context to the inherent curvature, the test objects are designed in a way that all properties can be well detected. So, the shape of the artificial sensor is cylindrical, resulting in a base to tip diameter ratio of $\vartheta = 1$.

- The overall shape of a vibrissa hair shaft is adapted to the artificial sensor shaft. The length dimension is chosen to be much larger than the diameter dimension d, resulting in a slender shape with a diameter of 0.5 mm and a length of 0.1 m. The slenderness contributes essentially to the elasticity of the system. The sensor shaft consists of a spring steel alloy (E, Re, ν, ρ). All values and properties are summarized in Table B.3. The choice regarding the values of the diameter and length is driven by the measuring range of the used sensors and evaluated due to a pre-experimental analysis, see Table B.2.

- In accordance to literature, it is assumed that the information about an object of interest is totally represented in the support reactions: F_x, F_y, and M_z. Thus, the FSC is replaced by a combination of a conventional force and torque sensor, see Table B.2. Consequently, all visco-elastic properties and adaption mechanisms of an FSC are neglected. This is necessary, in this first approach, because the adaption of the structure and functionality of an FSC is an ambitious intent that exceeds the scope of this work.

- The sensor hair shaft is clamped onto the sensors. The sensors are based on strain gauges and, consequently, must deform in order of an applied load. This deformation is small and neglected, resulting in the assumption that the support of the sensor shaft is rigid.

The assumption that all interactions take place in the x-y-plane restricts also the displacement of the sensor shaft. The vibrissa hair shaft is displaced along specific trajectories during whisking which depend on the particular scanning task and situation, see Subsection 2.1.2. Hence, the displacement of the artificial sensor influences the recorded signal as well. Across all tasks reported in Subsection 2.1.3, the relative displacement between vibrissa hair shaft and object can be seen as a translational one and/or a rotation. For example, during wall following the displacement of the vibrissa hair shaft is translational. In contrast, the typical exploration whisking is characterized by a rotation t of the vibrissa.

In both cases, the vibrissa hair shaft is displaced across the resting object. So, also in the artificial sensor concept, the support of the sensor is displaced and not the object.

- The sensor is displaced along a *straight* trajectory given by coordinates x_0 and y_0 with a constant, small velocity v.

A pre-analysis shows that a linear displacement corresponds to a constant stationary part of the recorded support reactions when a straight object is scanned, see Chapter A. Whereas a rotation given by a change of the angular position φ_0 causes a low-frequency change of the support reactions. If further unknown effects are analyzed, it seems to be advantageous when a straight, constant object shape causes a constant stationary signal to avoid difficulties in the analysis. Especially, when the influence of friction is analyzed as in Section 5.3 a translational displacement allows for a comprehensible interpretation of the results because the direction of the friction force can be directly predicted in the case of an object surface parallel to the scanning trajectory. As mentioned in Subsection 2.1.3, animals probably operate with respect to different coordinate frames a local vibrissa centered frame and a global head/animal centered one. The mentioned effect that a straight surface causes a constant stationary signal part depends on the definition of the frame corresponding to the support reactions. Here, the local frame located at the base of the sensor shaft is only linearly shifted away from the global frame. In case, a rotation of the local frame is enabled as well, also a rotatory actuation can cause a constant signal when a concave, circular shaped and coaxial oriented object is scanned. But, in a first step, the local frame is only shifted. Moreover, the vibrissa hair shaft is pro- and retracted in most cases. This is mirrored ín the artificial sensor concept.

> For the present work, it is defined that during the transition from pro- to retraction the sensor shaft does not release the contact with the object. Here, first, the tip of the sensor shaft gets pulled/protracted across the surface and afterwards pushed/retracted.

The described displacement is realized using two linear guides, see Table B.2. One linear guide positions the sensor in a defined vertical distance η to the object and the other one is used to realize the artificial scanning sweep (changing x_0 with respect to v).

The experiments with the artificial sensor are accompanied by theoretical simulations in order to identify and describe the inherent characteristics and effects of the artificial sensor, see Chapter A. In Figure 4.3, the three different representations of the sensor are compared. Summarizing, the actuation strategy of the present approach is a combination of the strategies observed for animals performing different scanning tasks. For example, the translational displacement of the sensor is inspired by wall following and the pro- and retraction phase by whisking.

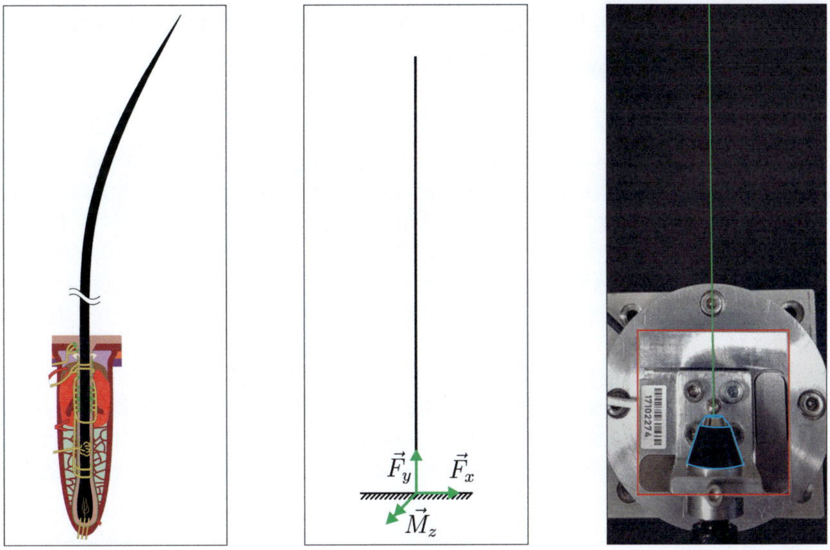

(a) Vibrissa **(b)** Theoretical model **(c)** Artificial sensor

Figure 4.3 – From natural paragon to artificial sensor: **(a)** vibrissa; **(b)** theoretical model of a vibrissa; **(c)** artificial sensor. In **(c)** the colors highlight the sensor shaft (green), the clamping (cyan), and the connected sensors (red), compare Table B.2.

The corresponding recorded support reactions can be interpreted as time series signals which can be used to determine, e.g., the object location at each moment and an integration across these contact locations results in a representation of the shape of the contacted object, see Section 5.1.

> Considering the huge simplifications, comparing the present approach to a natural vibrissa, the recorded signals may be of worse quality but the principal characteristics of the signals are kept. Since the focus of the present work is to analyze principal effects and not on the optimization of the sensor design, these restrictions seem to be applicable. But, the author of the present work is aware that the results generated with the artificial sensor must be very carefully interpreted when compared to natural vibrissae.

4.3 Provided information and definition of the sample properties

Next to the assumptions regarding the vibrissa/sensor, also the object and its properties must be discussed, see Figure 4.1. Following the discussion in Subsection 2.2.2, an object exhibits information of different orders of geometrical size. The object shape is superimposed by multiple macroscopic surface features (MACFs) and a microscopic surface texture (MICT). The shape and the MACFs cause a change of the contact force as a consequence of a change of the position of the contact point. In contrast, the features of a MICT sum up in a contact characteristic described by the theory of dry friction, see Subsection 2.2.2. Thus, a change in the MICT will result in a change of the present friction coefficient between surface and sensor shaft. Therefore, scanning a MICT, the contact force is changing due to sticking and sliding of the tip and not due to a spatial displacement as for the object shape or MACFs. Figure 4.4 illustrates the described relations of the different object properties.

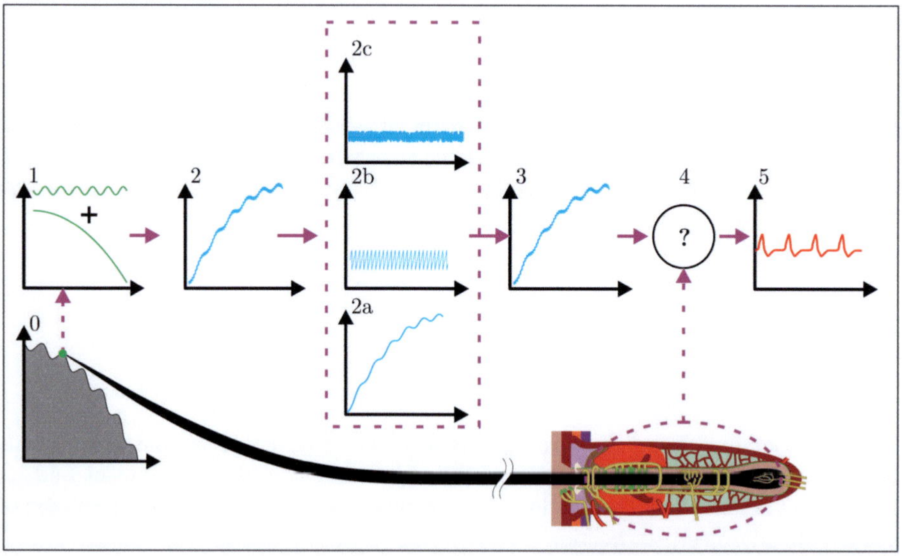

Figure 4.4 – Qualitative signal characteristics: 0- object shape superimposed by MACFs and a MICT; 1- spatial signal components of the object; 2- contact force; 2a- low-frequency signal components; 2b- high-frequency signals components/stick-slip effect; 2c- noise/deviations; 3. support reaction; 4- unknown signal transformation inside the FSC; 5- generated action potential.

The complete information about the object properties enters the vibrissa/artificial sensor through the contact point. The information gets transformed multiple times, see Subsection 2.2.1. The transformation from the support reactions inside the FSC to an action potential as the output of the mechanoreceptors is neglected because here the neurobiological periphery begins which is excluded from the present work. One objective of the present work focuses on the evaluation of the support reactions and the decomposition of the different information of the object. Nevertheless, the applied algorithms are roughly inspired by the characteristics of the mechanoreceptors. For example, the RA and SA mechanoreceptors distinguish and separate the low from the high frequency part of the captured signal. An alike filtering is necessary for the present approach, see Chapter 5.

The test objects must be designed in a way that the different properties of an object are recognizable with the present sensor concept:

- The test objects are supposed to be rigid and they exhibit the expected geometrical shape.
- The MICT is defined to be represented by features that are an order of magnitude smaller than the diameter of the sensor shaft and they are assumed to be stochastically distributed. These requirements are fulfilled by sandpapers with different grit sizes which also are used frequently in other works, see Table 2.1.
- The MACFs are defined to be essentially larger than the diameter of the tip in order to result in a recognizable spatial displacement of the contact point during scanning. For the present work, the MACFs are periodically distributed with a low spatial frequency.
- Finally, the geometrical dimension of the object shape is much larger than the one of the MACFs, and therefore, it causes a strong deflection of the sensor shaft.

For the experiments in Chapter 5, the test objects are designed in either way as a combination of the different information or to represent the information isolated. A detailed overview and information regarding the test objects are given in Section B.2.2.

5 Surface texture decomposition and feature extraction

This chapter provides the results of the performed experiments and simulations. In the next section, the relations between object shape and measured support reactions are analyzed. Subsequent, the object shape gets overlaid with MACFs, see Section 5.2. Afterwards, the influences of a MICT on the recorded support reactions are evaluated in Section 5.3. Finally, all findings are discussed and interpreted under aspects of technical application and in comparison to the natural paragon in Section 5.4. All used software tools, experimental equipment, and related properties are described in Chapter B. Furthermore, the essential processing of the measured data is explained in Section B.3. For both, experiment and simulation, it is defined that a scanning sweep starts with protraction of the sensor shaft, and can be followed by retraction. During protraction, the support is displaced in negative x-direction (from right to left) and the convex side of the deformed sensor shaft points in displacement direction, see Figure 5.1. Consequently, the retraction corresponds to a displacement in positive x-direction and the convex side of the deformed sensor shaft points forward. The scanning velocity is set to be $v = 0.001\mathrm{m\,s^{-1}}$. This choice is driven by the preliminary experiments, see Section B.4. In experiments, higher displacement velocities lead to vibrations excited by the motors that are recorded by the connected sensors and hamper a good analysis of the different signals. In the following sections and subsections, Figure 5.1 is modified in order to correspond to the respective scanning scenario.

When time series signals of experimental data are shown, always only one repetition out of five is illustrated, compare Section B.3.

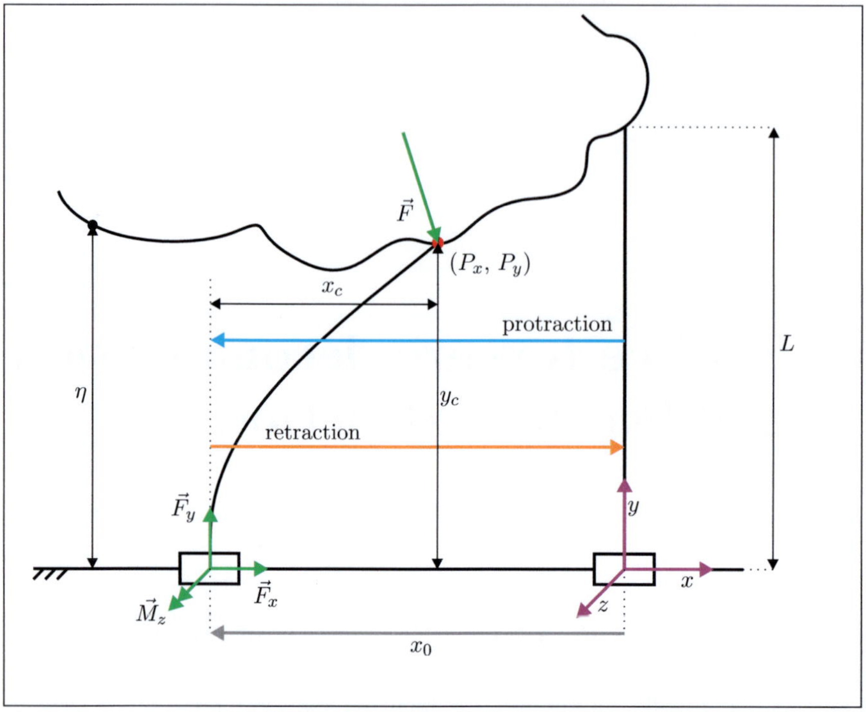

Figure 5.1 – The sketched scanning sweep is characterized by a protraction and retraction phase with a constant v and linear changing position x_0 of the clamping. The contact point location with respect to the actual clamping position is indicated by x_c and y_c. Using x_0, the contact point locations can be transformed into the object shape coordinates P_x and P_y which refer to the global frame.

5.1 Object shape

The relations between object shape and recorded support reactions at the base of a vibrissa-like tactile sensor are analyzed in detail in [Will; 2018]. Therefore, this section repeats known experiments and simulations and focuses on particular aspects with importance for the following discussion.

Three principle scenarios are analyzed. First, a horizontal flat contact plane is scanned for different η while pro- and retracting the sensor. Test object „s" represents the horizontal contact plane, see Table B.4. The same test object is used for the second scenario: the flat contact plane gets inclined by different angles α. Finally, the third scenario is the sinusoidal shaped test object f_1 is scanned, see Table B.4.

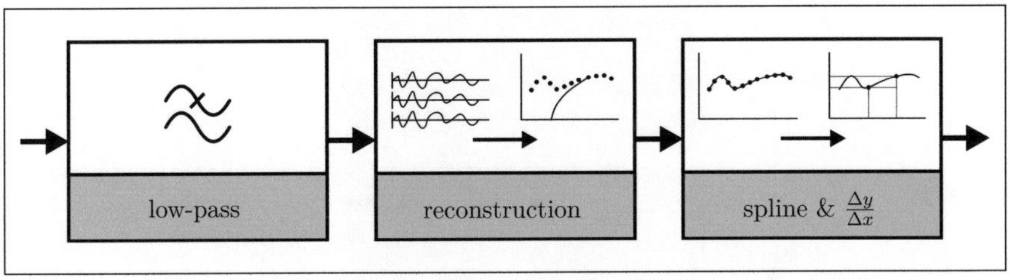

Figure 5.2 – Custom data processing in addition to Fig. B.2.

The data is processed in accordance to Section B.3, and the custom operations are explained in Figure 5.2:

- The recorded support reactions are filtered using a low-pass with a cut-off frequency of 5 Hz because it is supposed that the information belonging to the object shape is contained in the low-frequency part of the signals, see Figure 4.4. The filter characteristics are similar to the ones describes in Section B.3.
- Afterwards, the recorded support reactions are evaluated using the algorithm to determine the contact point location (P_x, P_y), compare Section A.1. This step is summarized in the *reconstruction* block.
- The reconstructed objects gets fitted with a smoothing spline using the MATLAB function `cspas(·)` to finally determine the slope of the object shape α by numerical differentiation (MATLAB function `diff(·)`.

5.1.1 Horizontal contact plane

The support is 0.3 m protracted and, afterwards, retracted while the sensor shaft touches the object, see Figure 5.3(a). The vertical distance η between sensor shaft support and object is incremented to evaluate the relations between η and the support reactions F_x, F_y, M_z. Eleven different distances are analyzed, whereby $\eta = 0.04\,\text{m}\,(0.005\,\text{m})\,0.09\,\text{m}$. Here, the expected η is equal to the expected P_y. The object surface is assumed to be smooth. Consequently, the frictional interaction between sensor shaft and touched surface is supposed to be low in experiments and, in simulations, friction is neglected. Figures 5.3(b) and (c) show the experimental setup as well as results of a simulation. The support reactions for different η are illustrated as time series signal in Figure 5.4. The assumption that a constant distance between displaced support and scanned object causes a constant low-frequency/stationary signal part (see Section 4.2) can be confirmed for the protraction phase and in parts also for the retraction phase. The trends of all three support reactions are smoother when the sensor shaft is protracted than as for retraction. Since the object is assumed to be frictionless, F_x should be zero, see Figures 5.4(a) and (d).

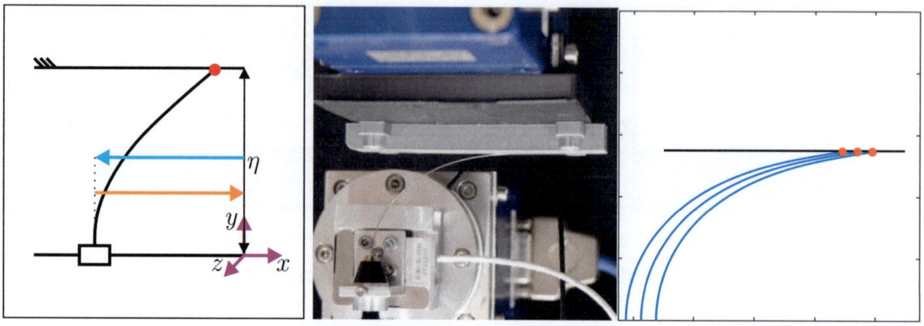

(a) Sketch **(b)** Experiment **(c)** Simulation

Figure 5.3 – Scanning test object „s": **(a)** sketch of the present scenario; **(b)** experimental setup; **(c)** simulation. In **(a)**, the cyan arrow indicates protraction and the orange one retraction. In **(c)** the red dots correspond to the current position of the contact point along the sensor shaft (colored in blue) and the black line to the expected object shape.

All three support reactions exhibit similar characteristics. For the protraction, the trends are smooth and their magnitude depends on η. In contrast, for retraction and large η, the trends are sawtooth-like, with a slower growth that is followed by a sudden, faster change of the signal. This effect can be observed for large η only, for small η the trends are smooth. Generally, in the transition from pro- to retraction, there are abrupt changes of the signals. The values of F_y change comparing pro- and retraction as well as the values of M_z do. For all η, the average values of the support reactions are summarized in Figure 5.5. As in Figure 5.4, the experimental values of F_x show nearly constant values that change in sign between pro- and retraction and in simulation F_x is equal zero as expected. For protraction, F_x is negative, which implies that a force in positive x-direction loads the sensor shaft, opposed to the displacement direction of the support. In the case of retraction, F_x becomes positive and, consequently, the force that loads the sensor shaft in x-direction must be negative, again opposed to the displacement direction of the support. This behavior is typical for a frictional force, see Subsection 2.2.2 and Section 5.3. Considering friction, also the changes of the values of F_y and M_z in Figure 5.4 can be explained. Neglecting friction, F_y and M_z must be constant, but if friction is considered the magnitude and orientation of the contact force change and, therefore, also the recorded support reactions are affected. Furthermore, for increasing η, F_y and M_z are principally decreasing. Only for large η and a retraction of the sensor shaft there are increasing values. These configurations correspond to the signals exhibiting a sawtooth-like trend. The scanned object shape is reconstructed using the recorded support reactions, see Figure 5.6(a). In either case, experiment or simulation, the object shapes are successfully reconstructed excepting the experimental results for $\eta = 0.09\,\text{m}$.

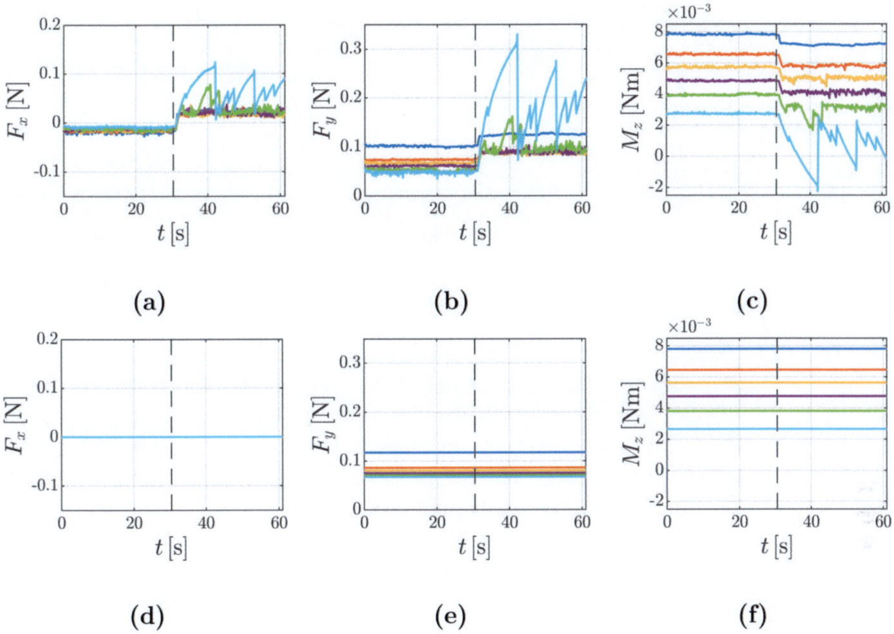

Figure 5.4 – The measured support reactions are illustrated in **(a)-(c)** and the simulated ones in **(d)-(f)**. The color indicates the distance η: blue- 0.04 m; red- 0.05 m; yellow- 0.06 m; magenta- 0.07 m; green- 0.08 m; cyan- 0.09 m. The black dashed vertical line indicates the reversing point in time of the support displacement.

The object shapes are pointwise, for each time step/sample, reconstructed and the results of the pro- and retraction are superimposed in Figure 5.6(a). Each entirety of points forms a straight horizontal profile. Object shapes that are reconstructed using the experimentally determined support reactions are located at a farther distance as expected. The object shapes belonging to the simulated support reactions are well aligned to the expected P_y, see Figure 5.7(a) and (b). The experimental results for $\eta = 0.09\,\mathrm{m}$ contain multiple outliers in the reconstructed shape, see Figure 5.6. Therefore, large deviations are observed in Figure 5.7. But, simultaneously, the corresponding object shape is well reconstructed as can be seen in Figure 5.6(a). This contradiction gets explained considering the findings of [Steigenberger et al.; 2015] and the limitations of the algorithm of [Scholz, Rahn; 2004], see Section A.1. The algorithm to determine the contact point location is based on the fact that the bending moment along the sensor shaft goes to zero at the contact point location on the sensor shaft. In [Hires et al.; 2013], [Steigenberger et al.; 2015], it is demonstrated that the sensor shaft can reach two different equilibrium states when its tip sticks to the surface. These states correspond to the first and second buckling mode. In the case of the second one, an additional inflection point is present along the deformed sensor shaft.

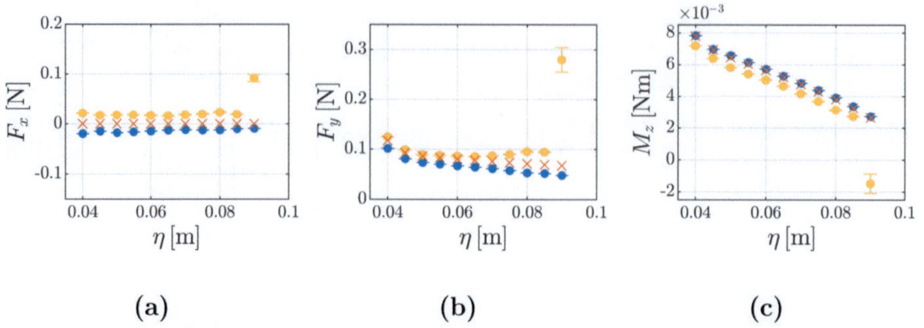

Figure 5.5 – The mean values and standard deviations of the support reactions for pro- and retraction are displayed with respect to different η. The experimental values correspond to blue (protraction) and yellow (retraction) dots. The simulated data are marked with red (protraction) and green (retraction) crosses. The red crosses are overlaying the green ones since the simulated data aligns perfectly.

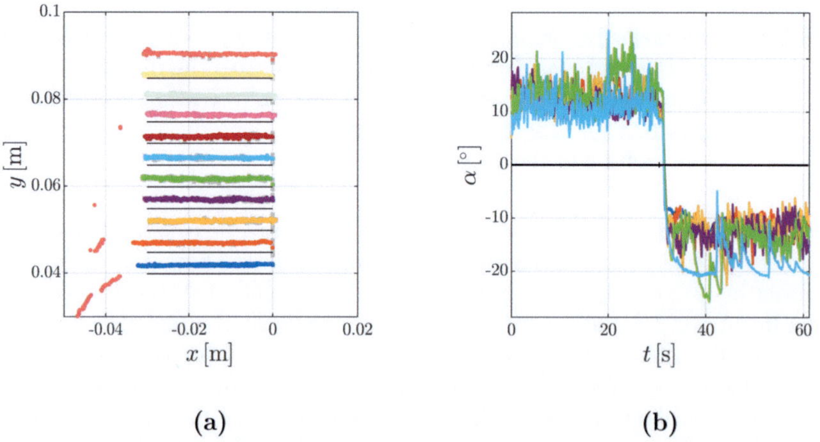

Figure 5.6 – In **(a)**, the reconstructed object shapes belonging to the experiments are colored in dependence on η: blue- 0.04 m; red- 0.045 m; dark yellow- 0.05 m; magenta- 0.055 m; green- 0.06 m; cyan- 0.065 m; dark red- 0.07 m; pink - 0.075 m; light grey- 0.08 m; light yellow- 0.085 m; rose- 0.09 m. The grey shade enveloping the reconstructed shapes correspond to the standard deviation of P_y. The results of the simulation are colored in black. In **(b)**, the color indicates the distance η: blue- 0.04 m; red- 0.05 m; yellow- 0.06 m; magenta- 0.07 m; green- 0.08 m; cyan- 0.09 m. The black vertical line corresponds to the simulations.

The presence of an inflection point along the deformed sensor shaft causes an S-shaped elastica exhibiting inflection points. This phenomenon is analyzed in context to vibrissa-like tactile sensors in [Steigenberger et al.; 2015] and in a more general context in [Singh, Goss; 2019]. Inflection points occur when the tip of the sensor shaft intensively sticks to the contact surface, see, e.g., Subsection 5.3.1, too.

The bending moment at this inflection point is also equal to zero and, consequently, the algorithm of [Scholz, Rahn; 2004] detects the position of the inflection point and not the position of the contact point. The described scenario takes place for extensive sticking of the sensor tip at the object surface and results in large deviations in the support reactions as shown in Figure 5.4. It can only appear when the sensor shaft is axially compressed, which is the case for retracting the sensor shaft in the present discussion. This behavior is only observable in the experimental results because in the simulation friction is neglected.

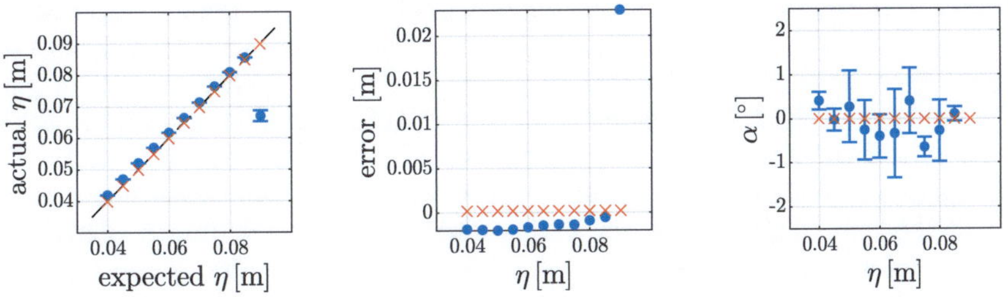

(a) (b) (c)

Figure 5.7 – The expected and actual object locations are compared in **(a)** and the corresponding error as difference between expected and actual value is given in **(b)**. In **(c)**, α is determined for each η. The value of α for $\eta = 0.09\,\mathrm{m}$ is exclude from the plot. In all subfigures, the blue dots correspond to the experimental data and the red crosses to the simulated one.

In Figure 5.6(b), the slope α of the object contour is determined by (A.7) using the support reactions F_x, F_y, and neglecting friction. For a horizontal straight contact surface it is expected that $\alpha = 0°$. This holds true for the simulations. But, for all experiments, the signals oscillate around $\alpha \approx 15°$ for protraction and $\alpha \approx -15°$ for retraction. Since α is equal zero, the found values must correspond to ζ following (A.7). This is confirmed by Figure 5.4(a) because there is a force in x-direction. Deriving α by numerical differentiation of the reconstructed object shape (compare Figure 5.2), α is determined correctly with respect to small deviations excepting the data belonging to $\eta = 0.09\,\mathrm{m}$, see Figure 5.7.

The results of this subsection show:

- Constant stimuli cause constant stationary signal parts.
- A smaller η leads to stronger support reactions.
- The higher-frequency parts of the signals seem to be without influence on the shape recognition.
- In experiments, even for a smooth surface, there is friction.
- The shapes of the test objects are successfully determined for all scanned η excepting $\eta = 0.09\,\mathrm{m}$.
- A given range for x_0 results in an equal range for the scanned section along the object.

5.1.2 Inclined contact plane

In a next step, the influence of an inclination of the scanned contact plane on the support reactions and derived information like the reconstructed shape and slope of the scanned object are analyzed. The contact plane is inclined by an angle $\alpha = 0°\,(5°)\,30°$. The upper edge of the contact plane is set to $\eta = 0.08\,\mathrm{m}$ for all α, see Figure 5.8(a). The support of the sensor shaft is 0.03 m pro- and retracted along the x-axis. The maximum of α is limited by restrictions of the experimental setup because for α larger than 30° the clamping of the sensor shaft collides with the test object. The object surface is still assumed to be smooth and in the simulation, friction is still neglected.

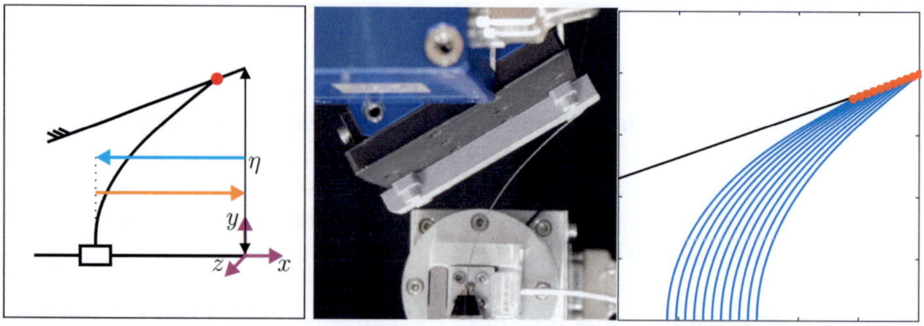

(a) Sketch **(b)** Experiment **(c)** Simulation

Figure 5.8 – Scanning the inclined ($\eta = 0.08\,\mathrm{m}$) test object „s“: **(a)** sketch of the present scenario; **(b)** experimental setup; **(c)** simulation. In **(a)**, the cyan arrow indicates protraction and the orange one retraction. In **(c)**, the red dots correspond to the current position of the contact point along the sensor shaft (colored in blue) and the black line to the expected object shape.

The support reactions for a complete scanning sweep are shown in Figure 5.9. Alike effects as for the horizontal contact plane in the previous subsection can be observed for the signals corresponding to $\alpha = 0°$. For example, while retracting the sensor shaft the recorded support reactions have again a sawtooth-like trend.

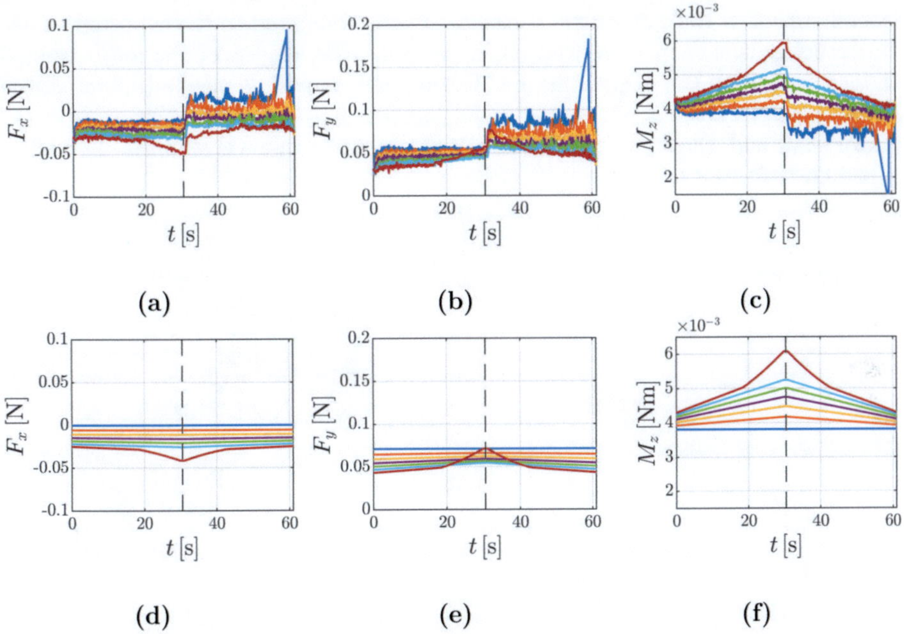

Figure 5.9 – The support reactions measured in experiment **(a)**-**(c)** and determined by simulations **(d)**-**(f)** are illustrated. In each subfigure, the different signals correspond to different α and are colored as follows: blue- $\alpha = 0°$; red- $\alpha = 5°$; dark yellow- $\alpha = 10°$; magenta- $\alpha = 15°$; green- $\alpha = 20°$; cyan- $\alpha = 25°$; dark red- $\alpha = 30°$. The black vertical dashed line indicates the transition from pro- to retraction.

In the transition from pro- to retraction, the support reactions change abruptly as before. But, for $\alpha \neq 0°$, there are differences compared to the support reactions measured scanning a horizontal plane. Here, F_x has a component that is caused by the orientation of the object shape and not by friction, see Figure A.2. This component of F_x is independent of the displacement direction. During pro- and retraction, the signals of F_x are decreasing with increasing α. For $\alpha = 30°$, the trend of the signal starts to diverge at $t \approx 20\,\mathrm{s}$. At this point, the contact point along the sensor shaft moves from the tip to a tangential location along the sensor shaft that causes a change of the contact force, compare Figure A.2. This can be also observed for F_y.

Furthermore, the trends of the signals of F_y are slightly inclined which is caused by the constant change of the vertical distance between the current contact location and support of the sensor shaft. In the case of M_z, these inclinations get more visible. The reconstructed object shapes are illustrated in Figure 5.10(a). The data belonging to the simulations aligns nearly perfectly with the expected values. In contrast, the alignment of the experimental results is poor compared to the expected values. Nevertheless, the shapes are well reconstructed. The lengths of the reconstructed shapes depend on the displacement of the support as well as on the contact status. For a tangential contact, the contact point location gets closer to the clamping of the sensor shaft and, therefore, a larger length of the object shape is scanned. This effect is demonstrated for $\alpha = 30°$.

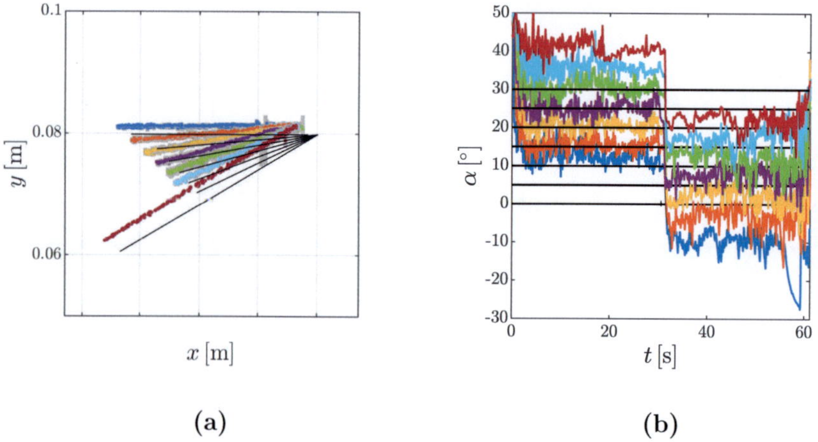

Figure 5.10 – In **(a)**, the reconstructed object shapes belonging to the experiments are colored in dependence on α: blue- $\alpha = 0°$; red- $\alpha = 5°$; dark yellow- $\alpha = 10°$; magenta- $\alpha = 15°$; green- $\alpha = 20°$; cyan- $\alpha = 25°$; dark red- $\alpha = 30°$. The grey shade enveloping the reconstructed shapes correspond to the standard deviation of P_y. The results of the simulation are colored in black. In **(b)**, the same color code is used.

The slopes of the object shapes are evaluated in Figure 5.10(b) by applying (A.7) and neglecting ζ. Using the simulated support reactions α can be determined well. Using the experimental data, the signals are approximately constant excepting the switch from pro- to retraction and small overlaying oscillations. In the case of protraction, the determined values of α are incremented by approx. 15° and for the retraction phase, the signals are decremented by the same offset. This offset is also found in Figure 5.7(b) and is identified as the influence of friction, which holds true for the present case, too.

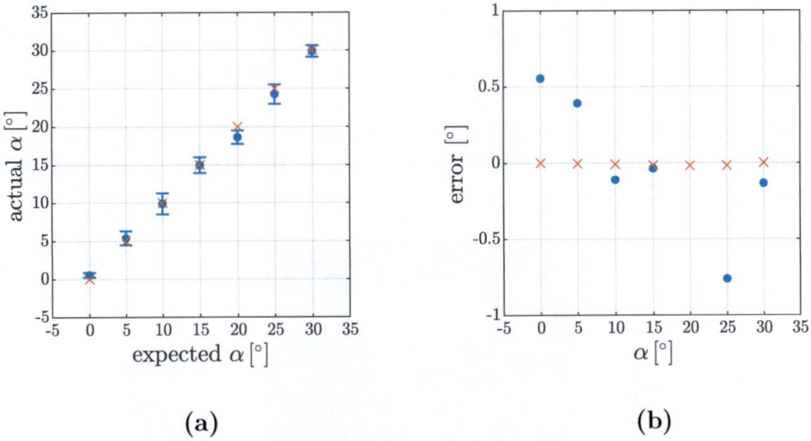

Figure 5.11 – In **(a)** the actual values of α are compared to the expected ones for the experimental and simulated data. The error in **(b)** is determined by the difference between the expected value and actual value. In both subfigures, the blue dots correspond to the experimental data, and the red crosses to the simulated one.

In Figure 5.11(a), the actual values of α are compared to the expected ones. The results of the simulations are better than the ones of the experiments, see Figure 5.11(b). But, in either case, the values are well determined. Considering the results shown in Figure 5.10, it must be highlighted again that the object shapes are reconstructed very well and only their position in the x-y-plane is displaced.

It can be summarized:

- The recorded support reactions change simultaneously to the change of the object shape.
- The object shapes are determined well, but their position in the x-y-plane is shifted.
- The size of the scanned section of the object shape depends on α.

5.1.3 Sinusoidal shape

The findings of the previous two subsections are combined and compared evaluating a more general object shape with the restriction that friction is still neglected. The scanned test object has a sinusoidal shape with convex and concave sections, see Figure 5.12 and table B.4. The test object is placed in two different distances from the clamping $\eta \in \{0.065\,\text{m}, 0.08\,\text{m}\}$. The right edge of the object serves as reference point, see Figure 5.12(a).

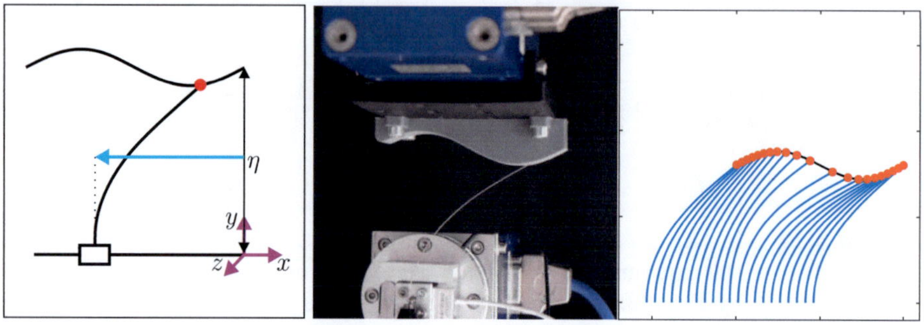

(a) Sketch (b) Experiment (c) Simulation

Figure 5.12 – Scanning the test object f_1: **(a)** sketch of the present scenario; **(b)** experimental setup; **(c)** simulation. In **(a)**, the cyan arrow indicates protraction. In **(c)** the red dots correspond to the current position of the contact point along the sensor shaft (colored in blue) and the black line to the expected object shape.

Due to the chosen object shape, the distance between contact point and clamping and the slope of the object shape changes continuously, which were kept constant in the previous investigations. A retraction of the sensor without detachment of the sensor shaft from the object seems to be counterproductive because of the object shape. Therefore, the object is scanned by only protracting the sensor. In Figure 5.13, the support reactions determined by experiments as well as simulations are compared for both levels of η. There are notable differences between the simulations and experiments. For F_x and F_y, the maximal values of the simulated signals are larger than those of the measured signals. Furthermore, the complete signals of F_x and F_y are shifted to slightly larger values compared to simulation. The trends of all signals are qualitatively alike but the simulated signals are more distinct. This can be also seen for the signals of M_z. In contrast to F_x and F_y, the simulated and measured signals of M_z align well, excepting the distinct minima in the simulated data. Focusing on the signals belonging to $\eta = 0.065\,\text{m}$, the signals change marginally at approx. $t \approx 22\,\text{s}$. Here, the contact changes from a tangential contact between sensor shaft and object shape to a contact between sensor tip and object shape.

These differences do not affect the reconstruction of the object shape, see Figure 5.14(a). In each case, the object shape is well reconstructed. The object shapes reconstructed using experimental data are marginally displaced in positive y-direction compared to the ones belonging to the simulations. For $\eta = 0.065\,\mathrm{m}$, the position in x-direction of the first reconstructed point is not zero because the sensor shaft was contacting the test object already along its shape in the starting position of the scanning sweep.

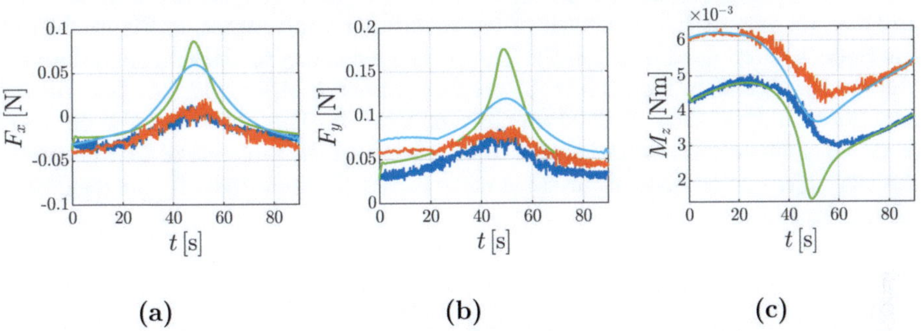

Figure 5.13 – In **(a)**, **(b)** and **(c)**, the support reactions F_x, F_y and M_z are illustrated for $\eta \in \{0.065\,\mathrm{m}, 0.08\,\mathrm{m}\}$. The blue (experiment) and green (simulation) curves correspond to $\eta = 0.08\,\mathrm{m}$ and the red (experiment) and cyan (simulation) ones to $\eta = 0.065\,\mathrm{m}$.

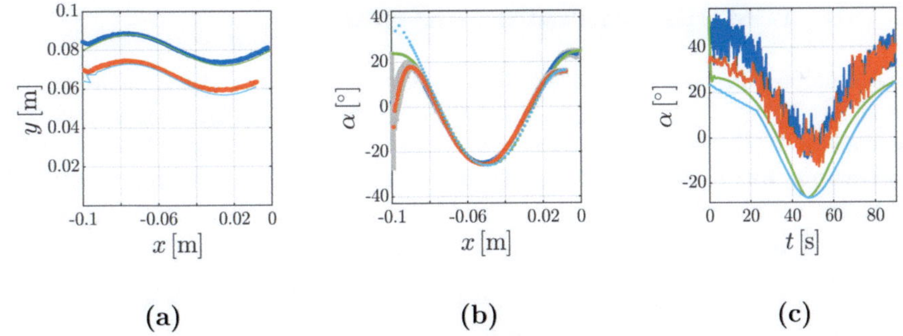

Figure 5.14 – In **(a)**, the reconstructed shape of test object f_1 is shown for both levels of η. The grey shade enveloping the reconstructed shapes correspond to the standard deviation of P_y. The slope of the reconstructed object shape is determined by numerical differentiation of the object shape in **(b)** and in **(c)** by applying (A.7) on the support reactions. The blue (experiment), green (simulation) curves correspond to $\eta = 0.08\,\mathrm{m}$ and red (experiment), cyan (simulation) ones to $\eta = 0.065\,\mathrm{m}$.

This situation is caused by the aim to align the data of experiments and simulations. Dropping this intention, a scan and reconstruction of the object shape including the mentioned section are possible. Moreover, also the reconstruction of the end-section of the test object are incomplete/incorrect in the case of the simulation results for $\eta = 0.065\,\mathrm{m}$. Since the second part of the test object exhibits a concave shape, there may be two contacts between the sensor shaft and object. This scenario is detailed analyzed in [Merker et al.; 2020]. For all four cases, these trends are nearly identical and deviate only at the boundaries from each other. In contrast, in Figure 5.14(c), the slope is determined using (A.7). The signals belonging to the simulated data are alike to the signals in Figure 5.14(b) because friction is still neglected. But, for the experimental data, there is again an offset of $\alpha \approx 15°$. This effect was also observed in previous investigations and seems to be related to the presence of friction. Since the test object „s“ consists of the same material like the test object f_1, the magnitude of the offset is alike, too.

It is demonstrated that:

- Complex object shapes can be analyzed in terms of a successful reconstruction of the object shape and determination of α.
- The findings for straight object shapes can be applied for more complex object shapes as well.

5.2 Macroscopic surface texture

The test objects representing MACFs are described in Table B.4. There are six different test objects f_2, g_1, g_2, g_3, h_1, h_2 and h_3, whereby g_2 and h_2 are identical. The overall shape of all objects is straight, excepting f_2 which has a sinusoidal profile as overall shape. The MACFs superimpose the overall shapes and are represented by sine functions with different frequencies and amplitudes. The post-processing of the experimental and simulated data is shown in Figure 5.15. The data is processed in three ways.

- First, like in Section 5.1, the support reactions are filtered by a low-pass with a cut-off frequency of 5 Hz.
- Following the first way (black arrow), the blocks and procedure are similar to Figure 5.2 in order to determine the reconstructed object shapes and slopes.
- The second (red arrow) and third (green arrow) way focus on the detection of the properties of the MACFs.
- In either case, the signals get filtered by a high-pass that excludes the low-frequency signal part.

- Afterwards, the peaks of the signals are evaluated by counting the peaks and determining the distances between them. Here, the MATLAB function `findpeaks(·)` is used. A peak is defined to be larger than the root mean square value of the corresponding signal.

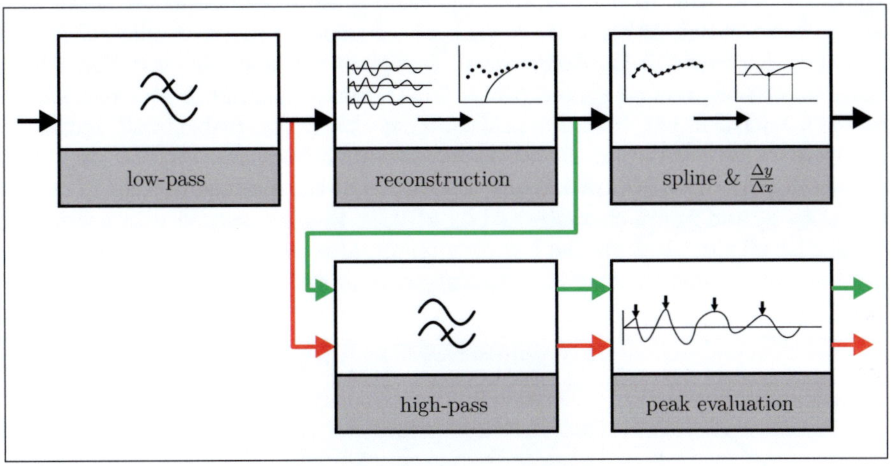

Figure 5.15 – Custom data processing in addition to Fig. B.2. The red data flow corresponds to the processing of the support reactions and the green one to the processing of the reconstructed object shapes. Both flows are independent of each other.

The MACFs are characterized by the properties: number of the peaks, distance between the features, height of the features, and their shape. The number of peaks and distance between the features can be determined by evaluating the recorded support reactions (second way/red arrow) as well as by evaluating the reconstructed object shape (third way/green arrow). But, the height and shape of a MACF can only be analyzed by evaluating the reconstructed object shape.

5.2.1 Horizontal contact plane with macroscopic surface texture

The test objects g_1, g_2, g_3, h_1, h_2 and h_3 are scanned for $\eta = 0.045\,\mathrm{m}\,(0.005\,\mathrm{m})\,0.09\,\mathrm{m}$, see Figure 5.16(a). The support is 0.03 m translationally displaced while the sensor shaft contacts the object. In contrast to Subsection 5.1.1, the sensor is protracted only because a retraction in the described way leads to the situation that the sensor tip gets stuck in the MACFs. This situation causes large support reactions, which are exceeding the maximal allowed values of the used sensor equipment.

The simulations are done without any influence of friction. The experimental setup and the simulation are illustrated in the Figures 5.16(b) and (c). Two observations are done inspecting the mentioned figures: it seems to be possible that the shaft contacts a test object at more than one point and that the contacts accumulate at the MACFs. The support reactions for $\eta \in \{0.045\,\mathrm{m}, 0.07\,\mathrm{m}\}$ are shown in the Figures 5.17 and 5.18. In all cases, the magnitude of the support reactions scales with the η as found in Subsection 5.1.1. A smaller η corresponds to larger values of F_y and M_z and smaller values of F_x. Furthermore, the signals belonging to $\eta = 0.07\,\mathrm{m}$ exhibit clearly observable, regular spaced peaks. This observation applies with restrictions also for $\eta = 0.045\,\mathrm{m}$ but, here, the peaks are not clearly observable in all signals. In Figures 5.17, for $\eta = 0.07\,\mathrm{m}$, the peaks have an alike frequency but the amplitude changes slightly. This is explained by the shapes of the scanned objects. The test objects g_1, g_2 and g_3 are characterized by MACFs that are spaced with a constant distance for all three test objects but their amplitudes are varied. The MACFs of g_1 have the smallest and the ones of g_3 the largest amplitudes.

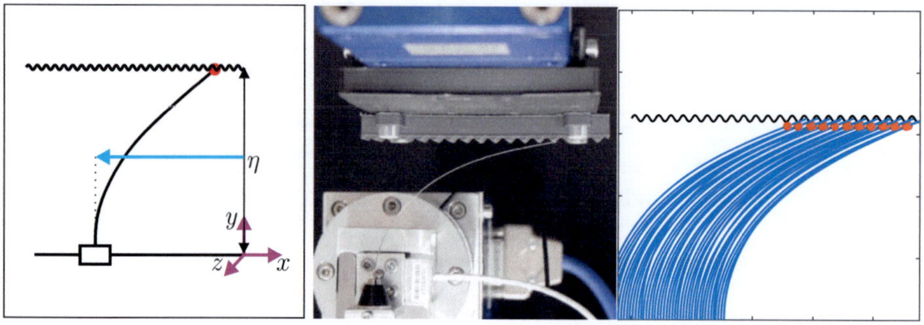

(a) Sketch **(b)** Experiment **(c)** Simulation

Figure 5.16 – Scanning test object g_2: **(a)** sketch of the present scenario; **(b)** experimental setup; **(c)** simulation. In **(a)**, the cyan arrow indicates protraction. In **(c)** the red dots correspond to the current position of the contact point along the sensor shaft (colored in blue) and the black line to the expected object shape.

However, the changes of the amplitudes of the peaks observed in the measured support reactions do not coincide with the variation of the amplitudes of the different test objects. The intuitive assumption is that a larger MACF leads to a larger peak in the support reactions. Nevertheless, in the present case, before a large peak occurs, a smaller peak in opposite direction can be observed. The smaller peak corresponds to the phase while the sensor shaft contacts a MACF and moves along its shape. When the sensor shaft reaches/passes the top of the MACF, it accelerates and bangs against the next MACF. This impact corresponds to the larger peak.

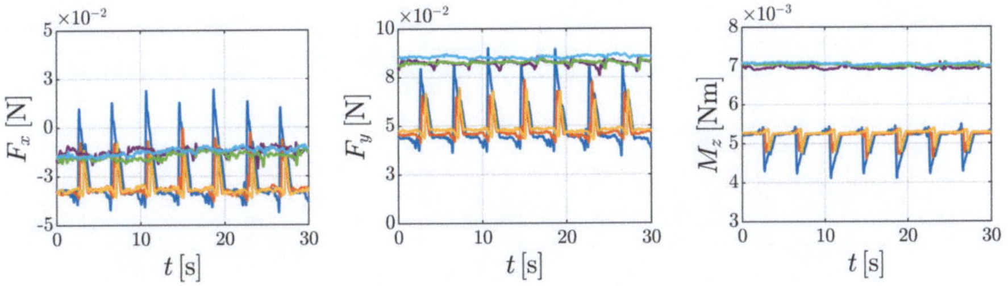

(a) (b) (c)

Figure 5.17 – All figures show experimental data only, the simulated data are alike. The shown support reactions correspond to: g_1, g_2 and g_3. The curves are colored in dependence on η and on the specified test object. Therefore, blue - g_1; red - g_2; dark yellow - g_3 for $\eta = 0.07\,\text{m}$ and magenta - g_1; green - g_2; cyan - g_3 for $\eta = 0.045\,\text{m}$.

Analyzing the data for $\eta = 0.045\,\text{m}$, the signals are characterized by an alike behavior as for $\eta = 0.07\,\text{m}$. But, the amplitudes of the peaks get smaller, in some cases not even observable. Furthermore, the peaks are shifted in time comparing the support reactions of both η. Figure 5.18 shows the data belonging to the test objects h_1, h_2 and h_3. The spatial frequency of the MACFs increased from test object h_1 to h_3 and the amplitude of the features remains constant for all three test objects. Analyzing the support reactions for $\eta = 0.07\,\text{m}$, the changes of the spatial frequency can be recognized in the frequency of the large peaks in the support reactions, best visible in Figure 5.18(c). But, as before in the case of the test objects of type „g“, the amplitudes of the peaks change in a non-intuitive way. The support reactions with the highest frequency exhibit the smallest amplitudes because the sensor shaft accelerates for a short time and, consequently, the following impact is less. In the case of $\eta = 0.07\,\text{m}$ and test object h_1, there are three closely following peaks. The first one corresponds to the moment when the sensor shaft reaches/passes the top of the MACF. The second one corresponds to the impact of the sensor tip at the bottom between the present and the following MACF. Finally, the third is related to the impact of the sensor shaft on the following MACF. This effect is possible because the distance between the MACFs is sufficiently large in relation to their size and the geometrical properties of the sensor shaft. For $\eta = 0.045\,\text{m}$, the described observations do not apply in full extent. As for the objects of type g, the peaks get smaller and the overall signal smoother. Generally, the findings of Subsection 5.1.1 regarding the influences of η on the support reactions are valid for the present results, too. A smaller η causes larger F_y and M_z, see Figure 5.5. However, in contrast, F_x is nearly constant in all cases also when friction is still neglected in simulation.

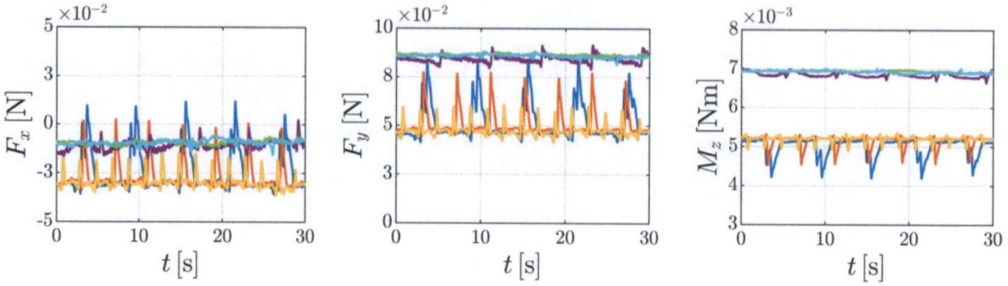

(a) **(b)** **(c)**

Figure 5.18 – All figures show experimental data only, the simulated data are alike. The shown support reactions correspond to: h_1, h_2 and h_3. The curves are colored in dependence on η and on the specified test object. Therefore, blue - h_1; red - h_2; dark yellow - h_3 for $\eta = 0.07\,\mathrm{m}$ and magenta - h_1; green - h_2; cyan - h_3 for $\eta = 0.045\,\mathrm{m}$.

Even if the overall object shape is still straight, the MACFs influence the deformation of the beam in the form that the contact force has a component in x-direction. Furthermore, due to the size of the MACFs and the geometric properties of the sensor shaft, it is possible that there are two contacts between sensor shaft and object. This scenario is detailed analyzed in [Merker et al.; 2020]. The influences of this effect are difficult to observe in experiments because they are present for a very short time only. But, in the case of simulation, they strongly affect the support reactions leading to large deviations. These deviations take place nearly simultaneously to the impacts of the sensor shaft on the MACFs. Therefore, both effects superimpose each other. But, the mean of the complete signal is affected by these deviations. Especially, the support reactions regarding the test object h_3 are influenced, see Figure 5.5(f). The reconstructed object shapes are shown in Figure 5.20. All reconstructed object shapes exhibit larger values of P_y as expected, with exception of the values the test object g_1. For larger η, the reconstructed shapes include the shapes of the MACFs. A larger distance between the features as well as a smaller amplitude of the features leads to a better result for the reconstructions of the shapes because these MACFs can be contacted well and they cause minimal dynamic effects. Larger amplitudes of the MACFs cause steep shapes and, therefore, the contact between sensor shaft and object tends to occur at the sides of the MACFs. An alike behavior is caused by smaller distances between the MACFs. For moderate η, frequently, multiple contacts between sensor shaft and test object emerge. For the present work, using the algorithm of [Scholz, Rahn; 2004] without further findings of [Merker et al.; 2020], these multiple contacts, between sensor shaft and test objects, lead to incorrect, predicted contact point locations and, consequently, to incorrect, reconstructed object shapes, see, e.g., Figure 5.20(f). Another effect is observable for the smallest analyzed η, here, there are nearly no MACFs are reconstructed.

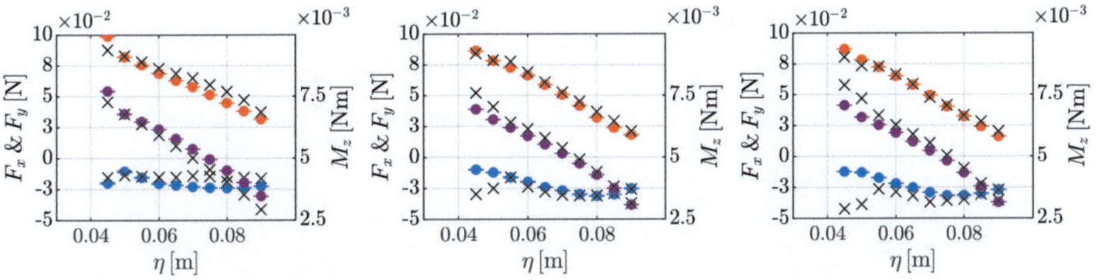

(a) g_1 **(b)** g_2 **(c)** g_3

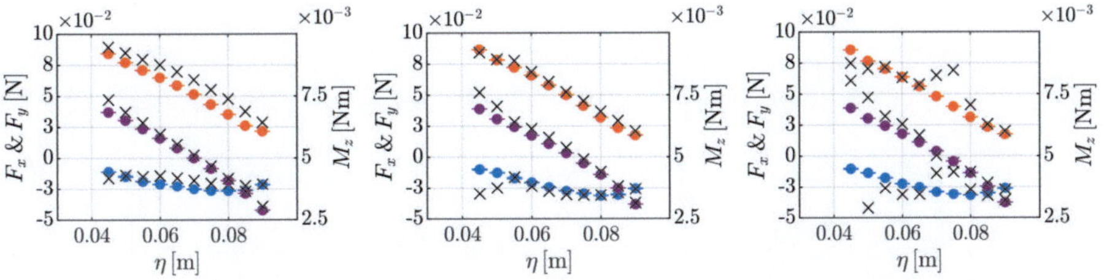

(d) h_1 **(e)** h_2 **(f)** h_3

Figure 5.19 – The figures **(a)**-**(f)** illustrate the means and standard deviations of the support reactions scanning the objects g_i to h_i, $i = 1, 2, 3$ for $\eta \in [0.045\,\mathrm{m}, 0.09\,\mathrm{m}]$ with: blue - F_x; red - F_y; magenta - M_z for the experimental data (dots), and the simulated data (black crosses).

The reconstructed object shapes are just straight because the bending induced curvature of the sensor shaft prevents that the tip of the sensor shaft enters the gaps between the MACFs. Evaluating the MACFs in detail, their number, distance and height are analyzed, see Figure 5.21. The number of the MACFs can be detected by analyzing the support reactions as well as the reconstructed object shapes. In the case of the support reactions, the number MACFs correspond to the number of peaks in the signals. In the case of the reconstructed object shapes, the number of the MACFs can be determined by the minimal values of P_y. Considering the displacement velocity, also the distance between the MACFs can be found by evaluating the support reactions by determining the time from peak to peak. In the case of the reconstructed object shapes, the distance between the features can be evaluated without additional information. The height of the MACFs cannot be detected using the support reactions. The height, as well as the information about the shape of the MACFs, is only contained in the reconstructed object shapes.

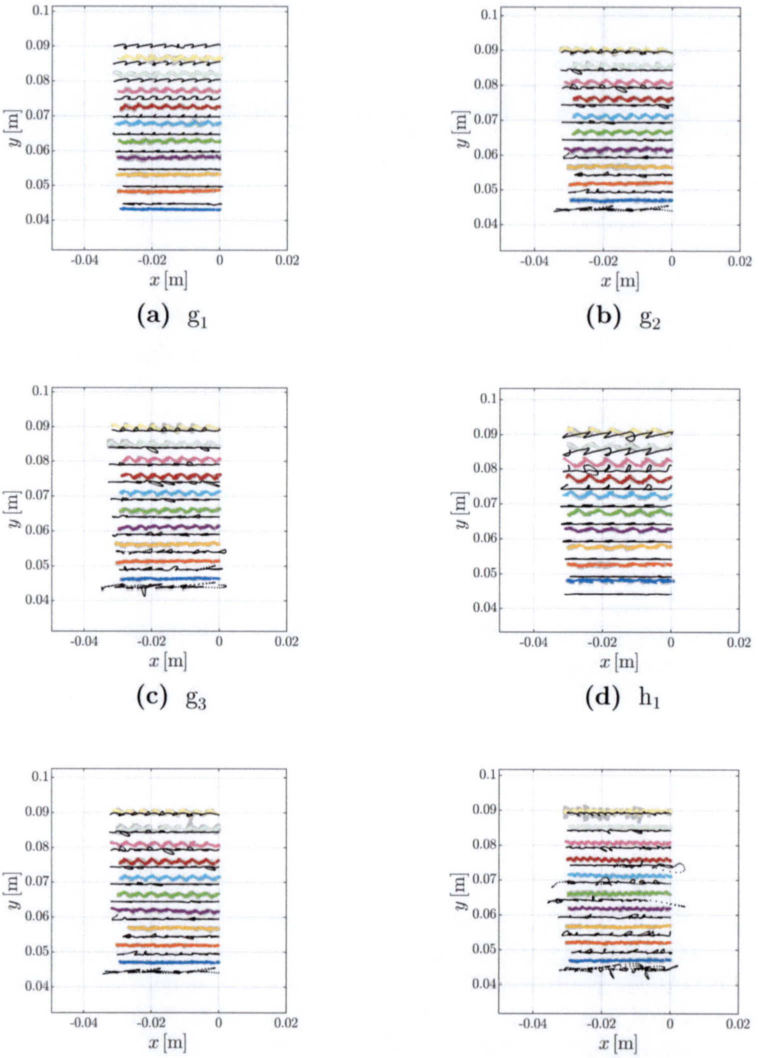

(a) g_1 (b) g_2 (c) g_3 (d) h_1 (e) h_2 (f) h_3

Figure 5.20 – The reconstructed object shapes are colored in dependence on η: blue- 0.045 m; red- 0.05 m; dark yellow- 0.055 m; magenta- 0.06 m; green- 0.065 m; cyan- 0.07 m; dark red- 0.075 m; pink - 0.08 m; light grey- 0.085 m; light yellow- 0.09 m. The grey shade enveloping the reconstructed shapes correspond to the standard deviation of P_y. The results of the simulation are colored in black.

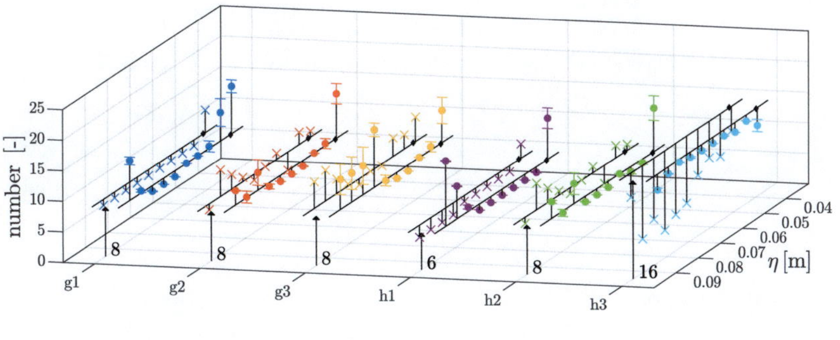

(a) Number of peaks detected in the recorded support reactions.

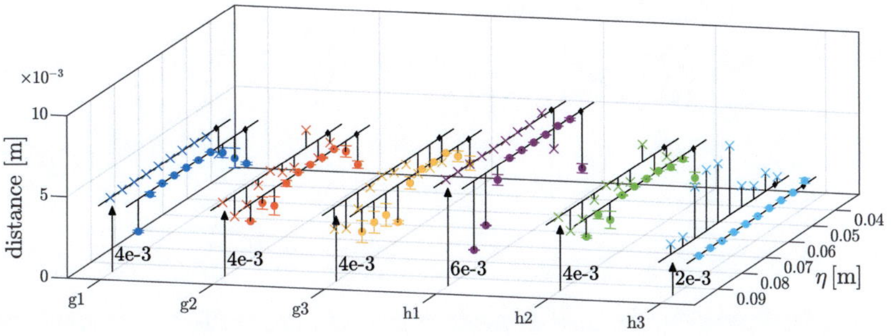

(b) Distance between the peaks detected in the recorded support reactions considering v.

Figure 5.21 – The determined properties of the MACFs are shown in: **(a)**-number; **(b)**-distances. The vertical arrows indicate the expected/real value of the regarding property. The dots correspond to experimental data and the crosses to simulated data. The diamond corresponds to the minimal value of η for the occurrence of a tangential contact [Steigenberger; 2013].

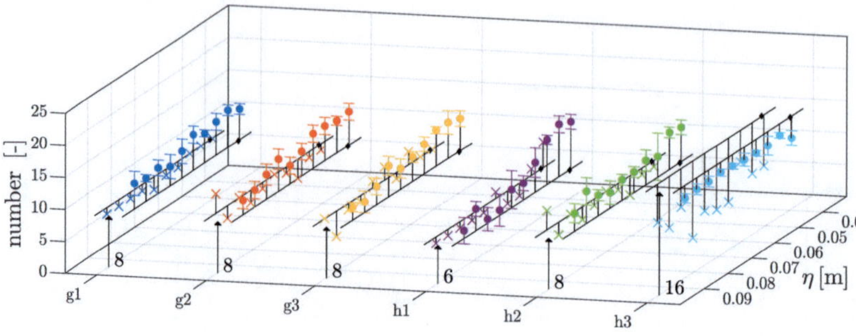

(a) Number of MACFs detected in the recognized object shapes.

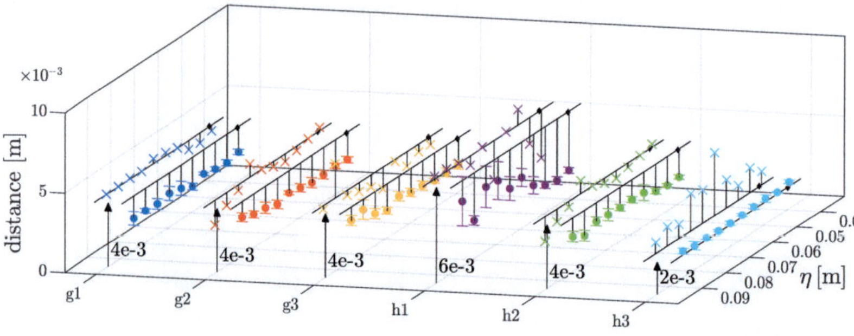

(b) Distance between the MACFs detected in the recognized object shapes.

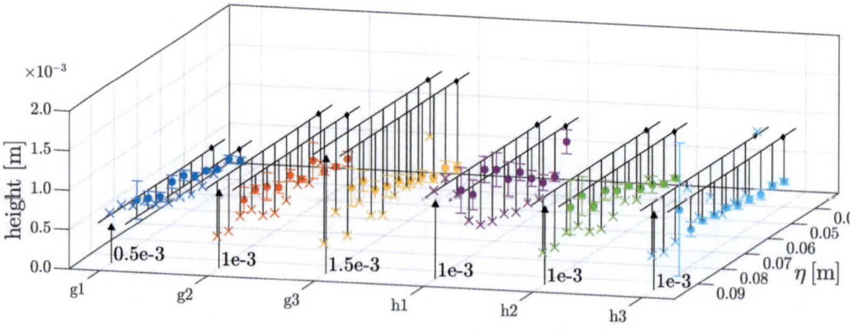

(c) Height of the MACFs detected in the recognized object shapes.

Figure 5.22 – The determined properties of the MACFs are shown in: **(a)**-number; **(b)**-distances; **(c)**-heights. The vertical arrows indicate the expected value of the regarding property. The dots correspond to experimental data and the crosses to simulated data. The diamond corresponds to the minimal value of η for the occurrence of a tangential contact [Steigenberger; 2013].

The expected and actual values of all mentioned properties are given in Figure 5.21. The number of the MACFs is well predicted for $\eta \in [0.05\,\mathrm{m}, 0.08\mathrm{m}]$ in most cases, see Figures 5.21(a) and (d). For larger and smaller η the results are worse. In some cases, e.g., the results for g_1 in Figure 5.21(a), all values are too large. Here, a systematic error occurs in the form that peaks are detected in starting and endpoint of the evaluated interval which in other cases count as one peak. Consequently, a constant error occurs. The distance between the features is better determined by evaluating the support reactions than by evaluating the reconstructed object shapes. Also here, for large and small η the results are erroneous. The used algorithm to detect the peaks in the signals can explain the difference in the results of the support reactions and the ones belonging to the reconstructed object shapes. The peaks in the support reactions are very steep and isolated. In contrast, the tops of the MACFs are approximately circular. Thus, the algorithm may detect the top too early or too late. The determination of the height of the tops is poor, excepting the values of g_1. This finding can be explained also by the limitations of the used algorithm. Furthermore, it seems to be related to the fact that also the distances to the reconstructed object shapes are determined with deviations.

The following observations can be concluded with respect to scanning objects with a straight overall shape superimposed by MACFs:

- The influences of η on the mean values of the support reactions are alike to Subsection 5.1.1.
- The presence of MACFs is recognizable by peaks in the support reactions.
- The object shape is encoded in the low-frequency signal part and the peaks superimpose this low-frequency signal.
- For larger η, the shapes of the test objects are determined well, indulging both the overall object shape and the smaller parts of the MACFs.
- For $\eta \leq 0.0457\,\mathrm{m}$ [Steigenberger; 2013], the parts of the object shapes belonging to the MACFs are not reconstructed. The bending induced curvature behaves like a morphological filter for the MACFs.

5.2.2 Sinusoidal shape with macroscopic surface texture

Superimposing the test object f_1 by MACFs result in the test object f_2, see Table B.4. The test object is scanned for $\eta \in \{0.065\,\mathrm{m}, 0.08\,\mathrm{m}\}$ by only protracting the sensor, see Figure 5.23(a). Experiment and simulation are illustrated in the Figure 5.23(b) and (c). Two observations done in Subsection 5.2.1 are already confirmed by this figure. In Figure 5.23(b), in the presented state, the contact between sensor shaft and object occurs along the sensor shaft and, therefore, the sensor tip does not enter the MACFs. In Figure 5.23(c), the contact points between the sensor shaft and object seems to accumulate at the MACFs again, they are not uniformly distributed.

The MACFs are represented by steep peaks superimposing the low-frequency part of the support reactions, see Figure 5.24. The experimentally determined support reactions align well with the ones determined by the simulations. The low-frequency signal parts are comparable to the ones in Figure 5.12. Comparing the support reactions for the test objects f_1 and f_2, it is obvious that the peaks contained in the support reactions belonging to f_2 are caused by the MACFs. Focusing on $t \in [0\,\mathrm{s},\, 20\,\mathrm{s}]$ in Figure 5.24(d)-(f), there are no clearly observable peaks in the experimental data and only one more or less irregular peak in the simulated signals.

(a) Sketch **(b)** Experiment **(c)** Simulation

Figure 5.23 – Scanning the test object f_2: **(a)** sketch of the present scenario; **(b)** experimental setup; **(c)** simulation. In **(a)**, the cyan arrow indicates protraction. In **(c)** the red dots correspond to the current position of the contact point along the sensor shaft (colored in blue) and the black line to the expected object shape.

In contrast in Figure 5.24(a)-(c), there are good visible peaks. During this period, the convex part of the object shape is scanned. As mentioned in context to Figure 5.23, the contact between sensor shaft and test object takes place along the sensor shaft for $\eta = 0.065\,\mathrm{m}$. Consequently, there is no acceleration phase and impact of the sensor shaft and no corresponding peak. In all cases, the distribution of the peaks is not uniform. For the periods $t \in [0\,\mathrm{s},\, 20\,\mathrm{s}]$ and $t \in [60\,\mathrm{s},\, 90\,\mathrm{s}]$, the number of the peaks seems to be lower than for the period $t \in [20\,\mathrm{s},\, 60\,\mathrm{s}]$. During this period, the first section of the concave object part is scanned. Here, the sensor shaft is pulled „away" from the object shape and not „pushed" against it as before while scanning the convex part. This effect is good visible in Figure 5.24. The reconstructed object shapes and slopes are shown in Figure 5.25. The overall sinusoidal shape is well represented in the reconstructed object shapes. Also, the MACFs can be partly recognized. For $\eta = 0.08\,\mathrm{m}$, they are reconstructed along the complete shape. In the case of $\eta = 0.065\,\mathrm{m}$, they are absent at the beginning and ending of the reconstruction. The curves of the slope of the reconstructed object shapes, in Figure 5.25(b), are smooth.

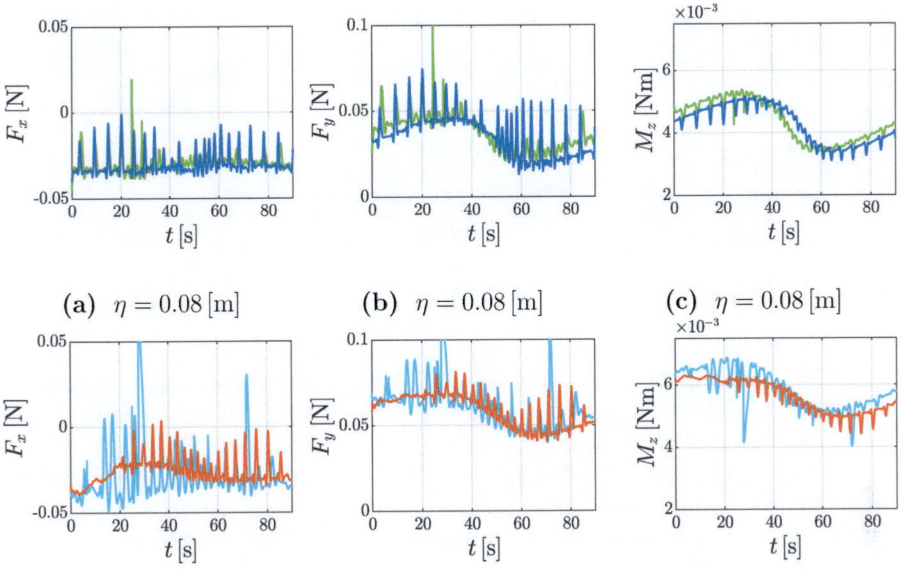

(a) $\eta = 0.08\,[\mathrm{m}]$ **(b)** $\eta = 0.08\,[\mathrm{m}]$ **(c)** $\eta = 0.08\,[\mathrm{m}]$

(d) $\eta = 0.065\,[\mathrm{m}]$ **(e)** $\eta = 0.065\,[\mathrm{m}]$ **(f)** $\eta = 0.065\,[\mathrm{m}]$

Figure 5.24 – In **(a)**-**(c)** and **(d)**-**(f)**, the support reactions F_x, F_y and M_z are illustrated for $\eta \in \{0.065\,\mathrm{m}, 0.08\,\mathrm{m}\}$. The blue (experiment) and green (simulation) curves correspond to $\eta = 0.08\,\mathrm{m}$, and the red (experiment) and cyan (simulation) ones to $\eta = 0.065\,\mathrm{m}$.

They lack the influence of the MACFs because identical coefficients for the spline as in Subsection 5.1.3 are used, compare Figure 5.15. Thus, the spline has the same effect as a low-pass filter and, consequently, the higher-frequency MACFs are filtered out and the slope of the overall object shape is determined. The results are alike to the ones in Figure 5.14(b). The slope of each object shape in Figure 5.25(c) is determined using the support reactions only. The peaks are related to the impacts of the sensor shaft on the MACFs during scanning. The results differ compared to the ones in Figure 5.25(b). The values are larger and the peaks observed already in the support reactions are also represented in the signals. Also a comparison to Figure 5.14(c) reveals further differences. In Figure 5.14(c) the determined object slopes are affected by the influence of friction. This holds true for the present signals as well but still, there is a difference between both results. Here, the slope of the reconstructed object shape belonging to $\eta = 0.08\,\mathrm{m}$ is shifted to a larger α and the signals for $\eta = 0.065\,\mathrm{m}$ to a smaller α. This effect gets not explained only by friction as before. Also here, friction influences the results, but this effect is caused by the presence of the MACFs. The calculated α depends on the support reactions and the value of the support reactions is related to the contact position between sensor shaft and object.

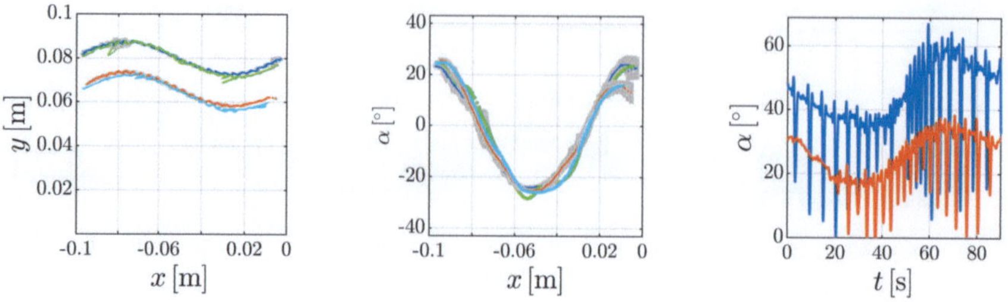

(a) (b) (c)

Figure 5.25 – In **(a)**, the reconstructed shape of test object f_2 is shown for both levels of η. The grey shade enveloping the reconstructed shapes correspond to the standard deviation of P_y. The slope of the reconstructed object shape is determined by numerical differentiation of the object shape in **(b)** and in **(c)** by applying (A.7) on the support reactions. The blue (experiment), green (simulation) curves correspond to $\eta = 0.08\,\mathrm{m}$ and red (experiment), cyan (simulation) ones to $\eta = 0.065\,\mathrm{m}$.

So, if the sensor shaft contacts a MACF at its side, the slope of the shape of the MACF is steeper than the one of the overall object shape, therefore, the determined α increases. Similarly, when the sensor shaft contacts the MACF at its top, the corresponding slope is may be smaller than the one of the overall object shape at this point. Hence, the determined object shape slopes depend on the position of the contact between sensor shaft and object/MACF. In the present examples, for a larger η, the sensor shaft enters well in the grooves between the MACFs and the contact point is located on its side and, as described, α increases. Here, it can be condensed reported that:

- The overall shape of test object f_2 is successfully determined for different η.
- The MACFs are partly recognized. The same effects as discussed in Subsection 5.2.1 are valid.
- The overall shape is encoded in the low-frequency signal part and the MACFs in the higher-frequency signal part.

5.3 Microscopic surface texture

The influence of friction between the test object and the sensor shaft is analyzed in this section. Therefor, the test objects „s“ and f_1 are covered with sandpapers with grit sizes from P3000 to P120 in order to represent different MICTs, see Tables B.4 and B.5. The sandpapers are glued with double-faced adhesive tape onto the test objects. The experiments are accompanied by simulations.

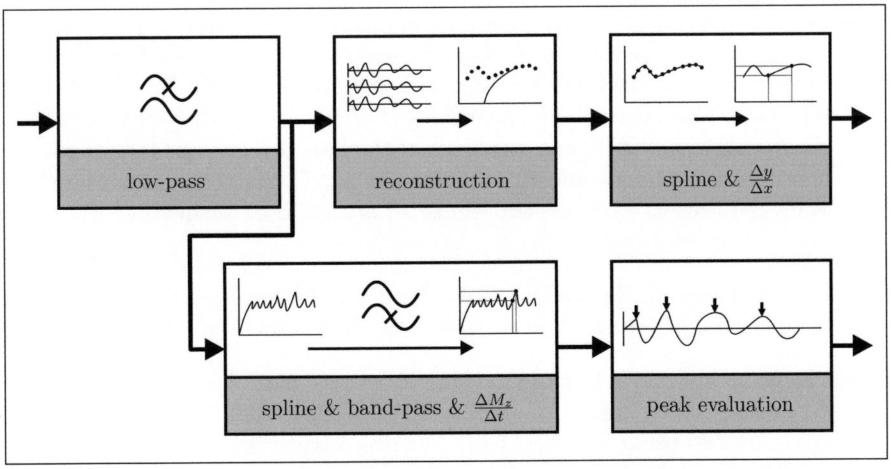

Figure 5.26 – Custom data processing in addition to Fig. B.2.

For the simulations, it is assumed that different MICTs are related to different friction coefficients. Hence, the dynamic friction coefficient is incremented $\mu_d = 0.1\,(0.1)\,0.8$ and the static friction coefficient is supposed to be $\mu_s = 1.2 \cdot \mu_d$. Thus, the maximal $\mu_s = 0.96$ and, consequently, the limits of the theory of dry friction are respected, see Subsection 2.2.2. The ratio $\mu_s/\mu_d = 1.2$ is chosen based on preliminary experiments and simulations. Horizontal straight test objects are scanned for different η and α in Subsection 5.3.1. The goal is to recognize the influences of the MICT on the captured support reactions. Especially, the differences between pro- and retraction of the sensor are focused whereby the sensor is displaced 0.02 m. Afterward, this approach is used equally to scan a sinusoidal shape text object, see Subsection 5.3.3. The corresponding data processing is shown in Figure 5.26.

- As in the previous sections, a low-pass filter with a cut-off frequency of 5 Hz is applied to all signals.
- The filtered signals are processed following two ways whereby both aim to evaluate the influence of the MICT.
- Following the first one, the signals are used to reconstruct the object shapes and to determine α. Using the determined α together with (A.7), the present μ_d can be calculated. To determine α, a smoothing spline is fit through the reconstructed object shape. This spline has the influence of a further low-pass filtering. So, higher-frequency signal parts caused by, e.g., the stick-slip effect, are excluded and, therefore, the determined friction coefficient corresponds to approximately to μ_d.

- In the second case, the support reactions are further analyzed. Here, the goal is to determine the number of stick-slip events that are encoded in the higher frequency signal part and to judge the present friction coefficients based on this value. This step follows the findings of [Wolfe et al.; 2008], [Schwarz; 2016], [Oladazimi et al.; 2018] and the ideas of the kinetic signature hypothesis/slip hypothesis/local code and can be done for F_x, F_y and M_z. Here, only M_z is evaluated because experimental signals of M_z are most adequate.

- The support reactions are filtered using a smoothing spline using the MATLAB function `cspas`(·). Subtracting this spline from the corresponding support reaction causes a high-pass filtering. This procedure is inspired by [Krystek; 2009].

- The numerical derivatives of the higher frequency part of the support reactions are determined to further emphasize the stick-slip events. The resulting signals are evaluated using the MATLAB function `findpeaks`(·) whereby a peak is defined to be larger than 50% of the mean of tenth-largest values of the time-series signal.

5.3.1 Horizontal contact plane with microscopic surface texture

The MICTs are systematically varied as well as $\eta = 0.04\,\mathrm{m}\,(0.01\,\mathrm{m})\,0.09\,\mathrm{m}$, see Figure 5.27(a) and Table B.5.

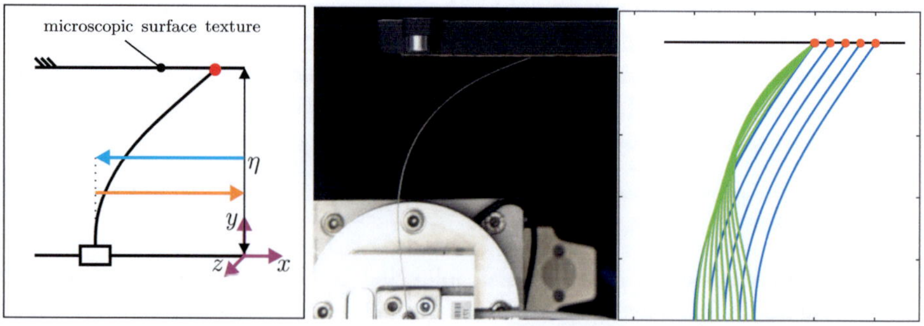

(a) Sketch **(a)** Experiment **(b)** Simulation

Figure 5.27 – Scanning test object „s“: **(a)** sketch of the present scenario; **(b)** experimental setup (P320), retracting the sensor; **(c)** simulation ($\mu_d = 0.8$). In **(a)**, the cyan arrow indicates protraction and the orange one retraction. In **(c)** the red dots correspond to the current position of the contact point along the sensor shaft (blue- protraction, green- retraction) and the horizontal black line to the expected object shape.

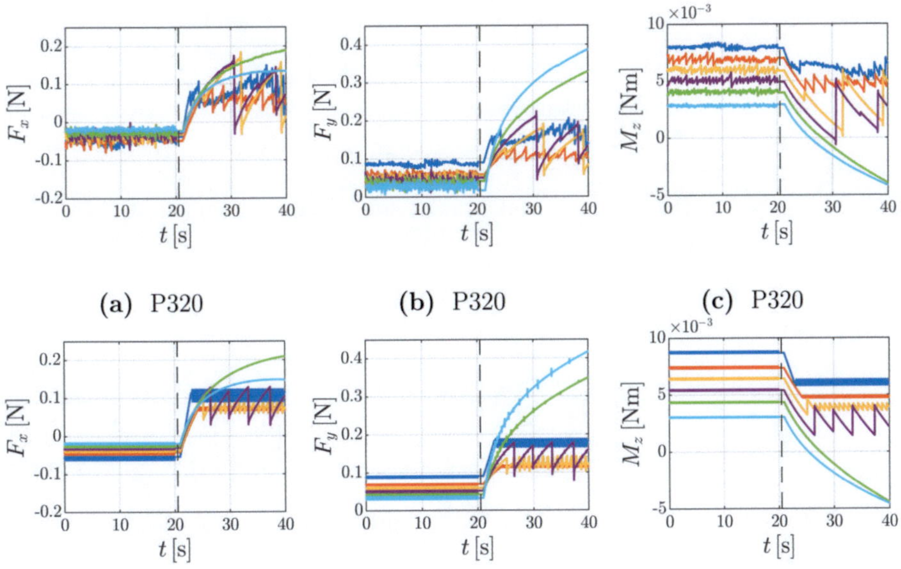

(a) P320 **(b)** P320 **(c)** P320

(d) $\mu_d = 0.6$ **(e)** $\mu_d = 0.6$ **(f)** $\mu_d = 0.6$

Figure 5.28 – The measured support reactions are illustrated in **(a)**-**(c)** and the support reactions determined by simulations in **(d)**-**(f)**. The color indicates the distance η: blue- 0.04 m; red- 0.05 m; yellow- 0.06 m; magenta- 0.07 m; green- 0.08 m; cyan- 0.09 m. The black dashed vertical line indicates the reversing point in time of the support displacement.

The elasticae during experiment and simulation are shown in Figures 5.27(b) and (c). For both cases, the elasticae for a retraction of the sensor shaft are illustrated. Compared to Figure 5.3, it is notable that there is an inflection point along the sensor shaft. This phenomenon is related to extensive sticking of the tip of the sensor shaft on the MICT as already discussed in Subsection 5.1.1 and in [Steigenberger et al.; 2015], [Scharff et al.; 2019a], too. As discussed before, this situation prevents a successful reconstruction of the object shape, compare Subsection 5.1.1. These states of extensive sticking can also be identified in the time-series signals of the support reactions, see Figure 5.28. Here, F_x, F_y are strongly increasing and M_z is strongly decreasing when η is large. For $\eta \in [0.08\,\text{m}, 0.09\,\text{m}]$, no abrupt signal changes can be observed like typical for the stick-slip effect. These changes become notable for $\eta \leq 0.07\,\text{m}$. Here, still, for a retraction of the sensor, the support reactions change moderately, followed by abrupt unloading. The amplitudes of these stick-slip events decrease and the number of the events increase for a decreasing η. Alike characteristics can be observed for protracting the sensor. Nevertheless, here, the amplitudes are smaller and the number of stick-slip events is larger than for retraction.

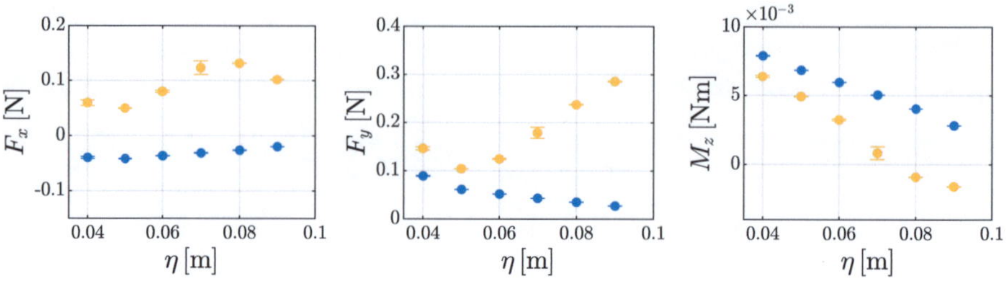

(a) P320 **(b)** P320 **(c)** P320

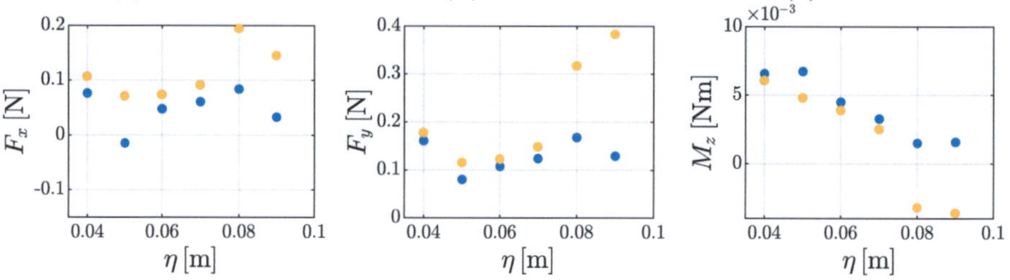

(d) $\mu_d = 0.6$ **(e)** $\mu_d = 0.6$ **(f)** $\mu_d = 0.6$

Figure 5.29 – The means and standard deviations of the support reactions are displayed with respect to different η. The experimental **(a)**-**(c)** and simulated **(d)**-**(f)** values corresponds to blue (protraction) and yellow (retraction) dots.

In Figure 5.29, the trend of the mean values of the support reactions are different to findings in Figure 5.5. The values of F_y strongly increase during retraction and for a large η. This is explained again by the extensive sticking of the tip of the sensor shaft on the MICT. The direction of the contact force tends to align to the y-axis and, consequently, F_y increases. Equally, the negative values for M_z in the cases of large η and retraction are explained. Excepting these values, the trends of the values for M_z are comparable to the ones in Figure 5.5. Analyzing the Figure 5.28(c) for $\eta \in \{0.06\,\mathrm{m}, 0.07\,\mathrm{m}\}$, the larger stick-slip events during retraction are disturbed by smaller ones while the sensor is sticking. These small superimposing stick-slip events may be so-called precursor events, which are caused by an unloading of the elastically deformed contact surface of the object [Rubinstein et al.; 2009]. This is probably because the sandpapers are glued by double-faced adhesive tape onto the test object.

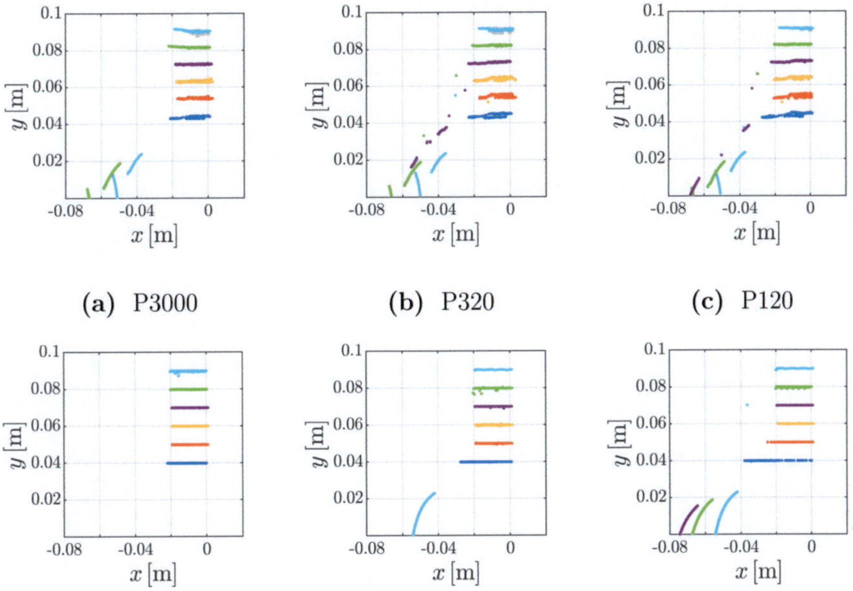

(a) P3000 (b) P320 (c) P120

(d) $\mu_d = 0.1$ (e) $\mu_d = 0.4$ (f) $\mu_d = 0.8$

Figure 5.30 – The reconstructed object shapes belonging to experiments are shown in **(a)**-**(c)** and the ones for the simulations in **(d)**-**(f)**. The reconstructed object shapes are colored in dependence on η: blue- 0.04 m; red- 0.05 m; dark yellow- 0.06 m; magenta- 0.07 m; green- 0.08 m; cyan- 0.09 m. The grey shade enveloping the reconstructed shapes correspond to the standard deviation of P_y.

The reconstructed object shapes are shown in Figure 5.30. They exhibit larger deviations than the ones in Figure 5.6. These deviations are caused by the MICTs because the stronger frictional interaction between object surface and sensor shaft leads to a stronger dynamic response of the sensor shaft as consequence of the periodic excitation. As discussed in context to Figure 5.28, for $\eta \in \{0.07\,\text{m}, 0.08\,\text{m}, 0.09\,\text{m}\}$ the elasticae exhibit an inflection point leading to false reconstructions during a protraction of the sensor. The reconstructed points for approx. $x < -0.04\,\text{m}$ are inflection points, see, e.g., Figure 5.30(c) and (f). For $\eta = 0.04\,\text{m}$, there is a tangential contact between sensor shaft and test object because $\eta \leq 0.0457\,\text{m}$ which is determined to be the necessary η for a tangential contact [Steigenberger; 2013]. Therefore, a larger section along the object shape is reconstructed in Figure 5.30.

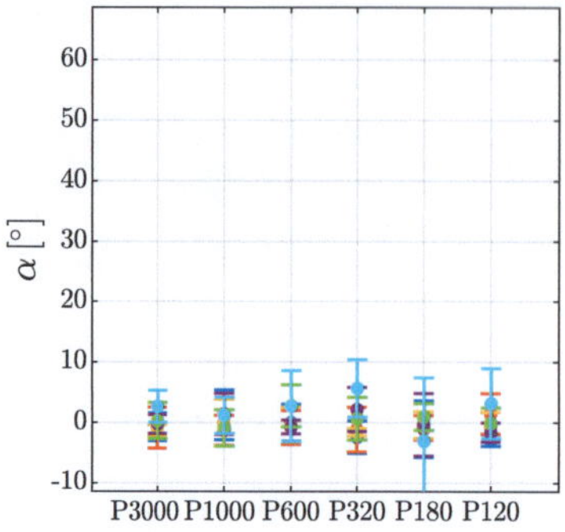

(a) Experiment/protraction

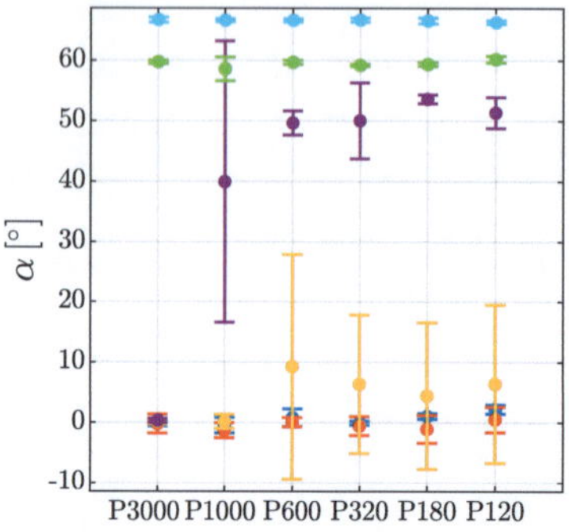

(b) Experiment/retraction

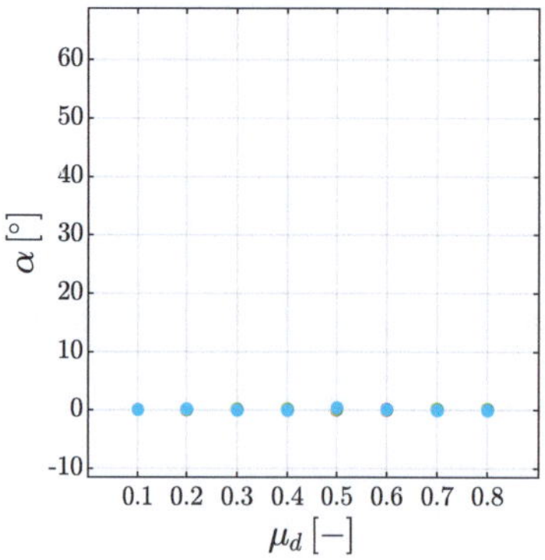

(c) Simulation/protraction

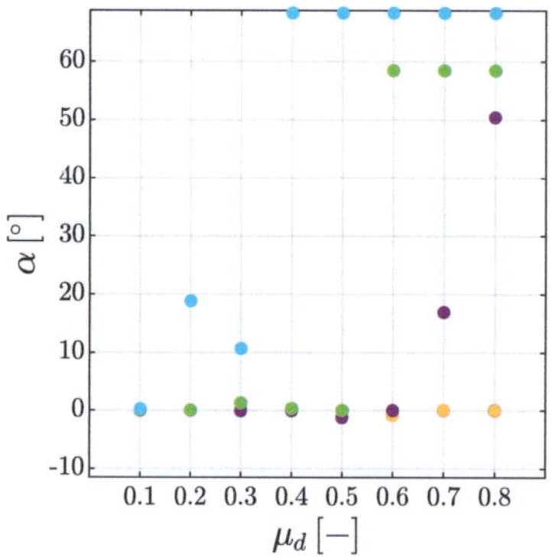

(d) Simulation/retraction

Figure 5.31 – The values of α for all types of MICTs and η are shown with: blue- 0.04 m; red- 0.05 m; dark yellow- 0.06 m; magenta- 0.07 m; green- 0.08 m; cyan- 0.09 m.

The corresponding α are given in Figure 5.31. In experiment and simulation, for a protraction of the sensor, α is close to zero as expected across all evaluated MICTs whereby the simulated data show nearly perfect results. In contrast, for retraction, the values of α are worse for large η. Since α is determined using the reconstructed shapes, the occurrence of the inflection points for large η are responsible for these deviations. The other α are well determined. With the information about the present α, the present μ_d can be determined using (A.7). The results are shown in Figure 5.32 for experiments and simulation. In experiments, the conditions in the contact between MICT and sensor shaft are not known a priori.

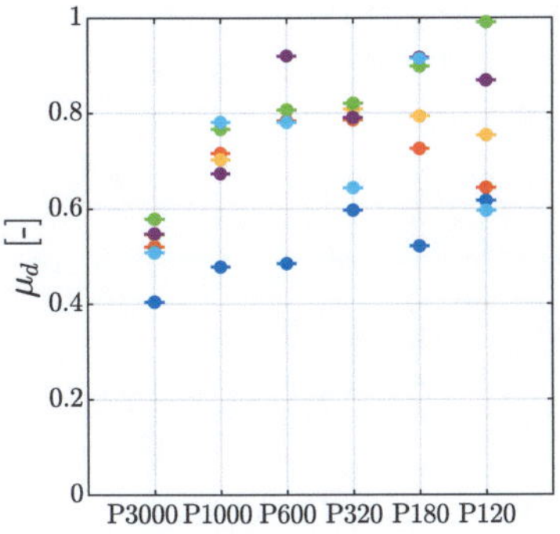

(a) Experiment/protraction

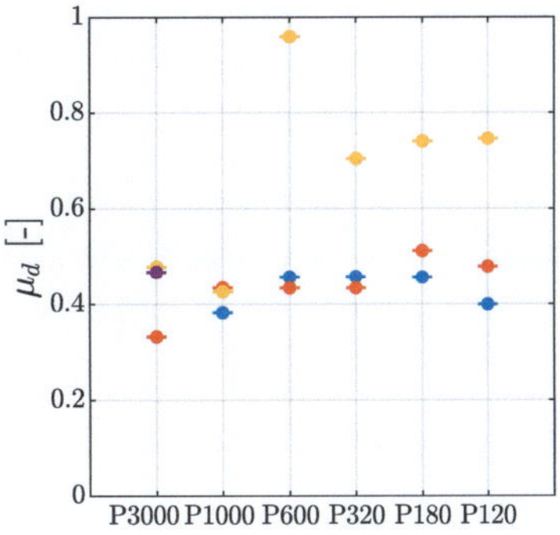

(b) Experiment/retraction

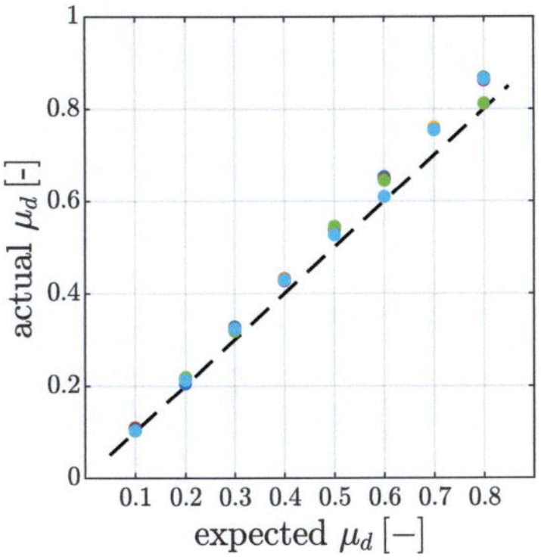

(c) Simulation/protraction

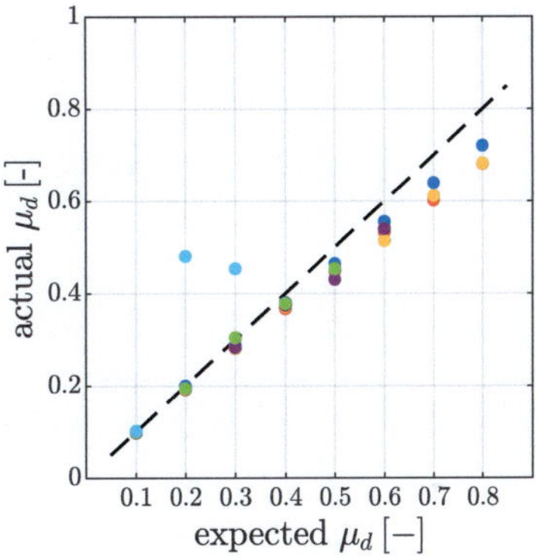

(d) Simulation/retraction

Figure 5.32 – The values of μ_d for all types of MICTs and η are shown with: blue- 0.04 m; red- 0.05 m; dark yellow- 0.06 m; magenta- 0.07 m; green- 0.08 m; cyan- 0.09 m. The dashed black line corresponds to the expected values. In **(b)**, all values for $\eta \in \{0.07\,\text{m},\ 0.09\text{m}\}$ are missing because they are larger than one and exceed the limits of dry friction, exepting the value for $\eta = 0.07\,\text{m}$ and the sandpaper P3000. Similarly, in **(d)**, the values of $\eta = 0.09\,\text{m}$ for $\mu_d \in [0.4,\ 0.8]$ and $\eta = 0.08\,\text{m}$ for$\mu_d \in [0.6,\ 0.8]$ are missing.

It is known that a larger average particle diameter corresponds to a smaller grit size and vice versa, see Table B.5. But, as discussed in Subsection 2.2.2, a larger average particle diameter is not necessarily related to a larger μ_d and/or μ_s.

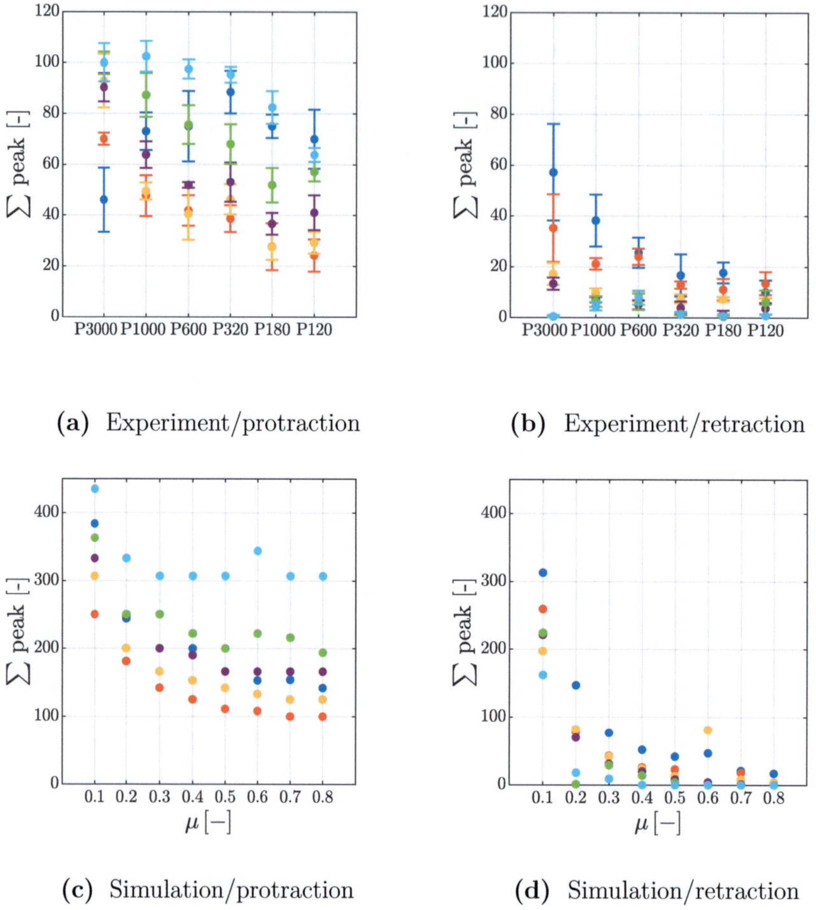

(a) Experiment/protraction

(b) Experiment/retraction

(c) Simulation/protraction

(d) Simulation/retraction

Figure 5.33 – The number of detected peaks of M_z for all types of MICTs and η are shown. The colored dots indicate the value of η: blue- 0.04 m; red- 0.05 m; dark yellow- 0.06 m; magenta- 0.07 m; green- 0.08 m; cyan- 0.09 m.

The determined μ_s in Figure 5.32(a) are roughly increasing from a sandpaper of type P3000 to one of P120. In the cases of P3000 and P1000, the determined values are at $\mu_d \approx 0.55$ and $\mu_d \approx 0.75$ for $\eta \in [0.05\,\text{m}, 0.09\,\text{m}]$ excepting $\eta = 0.04\,\text{m}$. Here, there is a tangential contact (compare Figure A.2) between the test object and sensor shaft. The values are smaller than the ones belonging to larger η. Generally, the values for μ_d are strongly changing for the same sandpaper and different η. This holds true for the values determined during protraction, see Figure 5.32.

The values of μ_d for large η and the sandpapers with small grit size are leaving the limits of the theory of dry friction because the used detection algorithm requires a well predicted α which is not the case for the mentioned configuration. In contrast, in simulation, the actual values match approximately the expected ones. For protraction, for large μ_d the actual values tend to be slightly larger than expected, but they are still well detected. In the case of retraction, the values for large η and large μ_d exceed the limits of dry friction because of a poor estimation of α, see Figure 5.32(b). In contrast to protraction, large μ_d tend to be smaller than expected if indeed the detection was successful. Inspired by the findings in Subsection 2.1.3, the number of stick-slip events is evaluated considering a given sensor displacement, see Figure 5.33. In experiment, it can be observed that a larger grit size causes a larger number of detected peaks. Furthermore, the number of peaks depends on η. For protraction, the number of peaks decrease with decreasing η, excepting $\eta = 0.04\,\mathrm{m}$. Here, also larger number of peaks are found because the measured support reactions get *noisy* and the detection algorithm fails in parts. The general change of the number of peaks can be explained by the influence of η. The stick-slip effect depends on μ_s, μ_d, v and F_n among others. If μ_s, μ_d and v are constant, only a change of F_n is possible. In the present case, F_n scales with η in a way that a smaller η a larger F_n. Consequently, the tip sticks longer on the surface and the number of stick-slip events decreases for a fixed scanning range. For retraction, there are larger numbers of peaks for small η which is in contrast to protraction. For smaller η, the authors of [Steigenberger et al.; 2015] show that when the sensor tip is pinned to one point and there is no further restriction due to any contact geometry, the elastica undergoes larger deflections until the slope of the elastica at the tip of the sensor shaft gets negative, see [Steigenberger et al.; 2015, Figure 10, pp. 12]. Interpreting this effect in the present context, the contact between sensor tip and test object gets opened and the tip starts sliding, even if the maximal μ_s is still not reached. Therefore, the determined number of peaks for $\eta = 0.04\,\mathrm{m}$ increases. This effect affects also other η but because of the tangential contact, this effect gets even forced. For scanning a test object with a distinct MICT, it is found:

- The object shape and α can be reconstructed by evaluating the low-frequency signal part during protraction.
- Based on the identified α, the present μ_d can be well predicted for moderate η.
- The number of peaks in the high-frequency signal parts depend on η, μ_d and μ_s.
- The signal characteristics between pro- and retraction are different and depend on η, a large η can cause problems during evaluation.
- Protracting the sensor causes low-frequency signal parts superimposed with high-frequency sawtooth-like oscillations with small amplitudes.
- Retracting the sensor causes low-frequency signal parts superimposed with low-frequency sawtooth-like oscillations with larger amplitudes.

5.3.2 Inclined contact plane with microscopic surface texture

The inclination of the test object „s“ is changed by varying $\alpha = 0° \, (5°) \, 30°$ to evaluate the relation between α and the different MICTs. The upper edge of the test object serves as reference point with $\eta = 0.08\,\mathrm{m}$, see Figure 5.34(a).

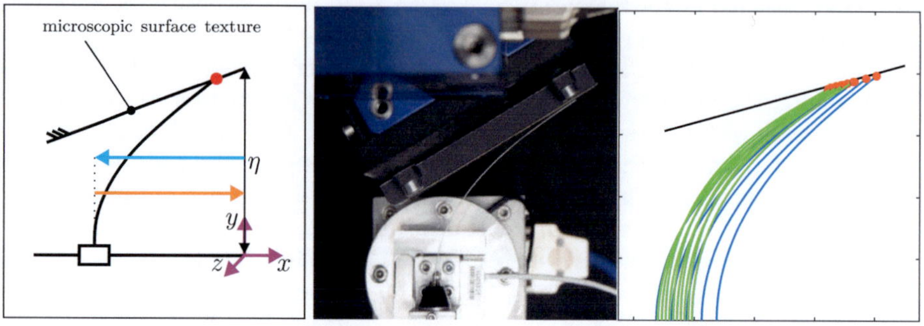

(a) Sketch (b) Experiment (c) Simulation

Figure 5.34 – Scanning test object „s“: **(a)** sketch of the present scenario; **(b)** experimental setup (P320), retracting the sensor; **(c)** simulation ($\mu_d = 0.8$). In **(a)**, the cyan arrow indicates protraction and the orange one retraction. In **(c)** the red dots correspond to the current position of the contact point along the sensor shaft (blue- protraction, green- retraction) and the black line to the expected object shape.

In Figure 5.34, it is notable that alike effects as observed in Subsection 5.3.1 occur when the test object is inclined by α. For a retraction of the sensor, the corresponding elasticae still exhibit inflection points, see Figure 5.34(c). The measured support reactions are given for the case of a sandpaper of type P320 and for every α, see Figure 5.35. There are low-frequency signal parts that are superimposed by a higher-frequency sawtooth-like trend when $\alpha \neq 0°$. At the beginning of the protraction phase, the experimental data shows various outliers, especially the data for $\alpha \in [5°, 15°]$ seems to be affected. Compared to the data belonging to the simulations, the trends of the support reactions are very alike. In Figure 5.9, the trends of M_z are clearly distinguishable due to the influence of friction the separation between the signals shrink and, consequently, they are less distinct in the Figures 5.35(c) and (f). This holds true for the experimental and simulation data. This effect influences also the reconstructed shapes, see Figure 5.36.

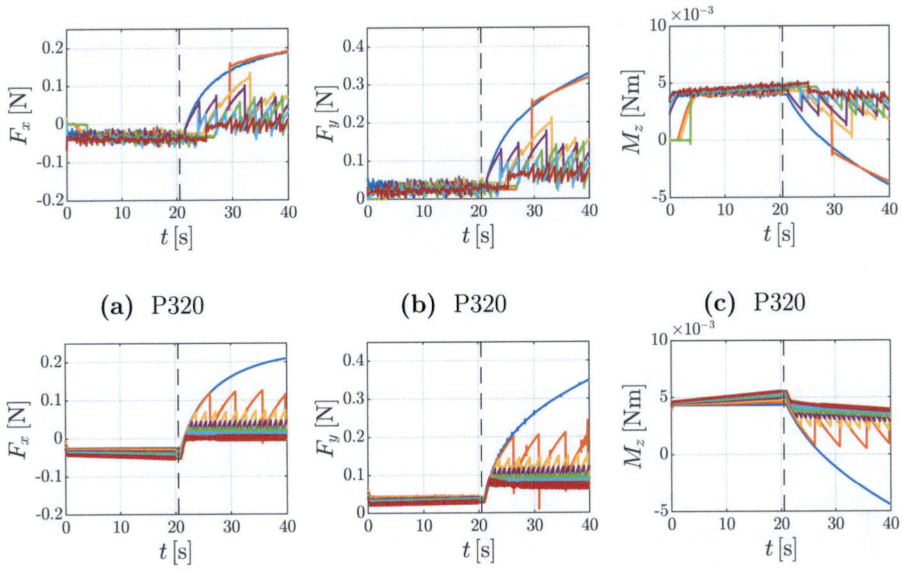

(a) P320 **(b)** P320 **(c)** P320

(d) $\mu_d = 0.6$ **(e)** $\mu_d = 0.6$ **(f)** $\mu_d = 0.6$

Figure 5.35 – The measured support reactions are illustrated in **(a)**-**(c)** and the support reactions determined by simulations in **(d)**-**(f)**. The color indicates the distance η: blue- 0°; red- 5°; yellow- 10°; magenta- 15°; green- 20°; cyan- 25°; dark red- 30°. The black dashed vertical line indicates the reversing point in time of the support displacement.

Here, in experiments, the object shapes are poorly reconstructed. Especially, in the case of P320, there are large deviations. In contrast, the results using the simulation data are acceptable. Already in Subsection 5.1.2, the results of the experimental data are of low accuracy. In the present case, they are even worse. A probable reason for this effect is that the differences between the support reactions for different α are very small compared to, e.g., different η, see Figures 5.4 and 5.28. So, deviations and *noise* included in the experimental data seem to be of an order of magnitude that is critical for the present reconstruction task. Since the simulation data is of much higher accuracy than the experimental one, the reconstructed object shapes are of better quality. The poor quality of the reconstructed shapes of the test object in the experiment influences strongly the determined values of α, see Figure 5.37(a). The trend of the determined values is principally increasing as expected but the values are far too large.

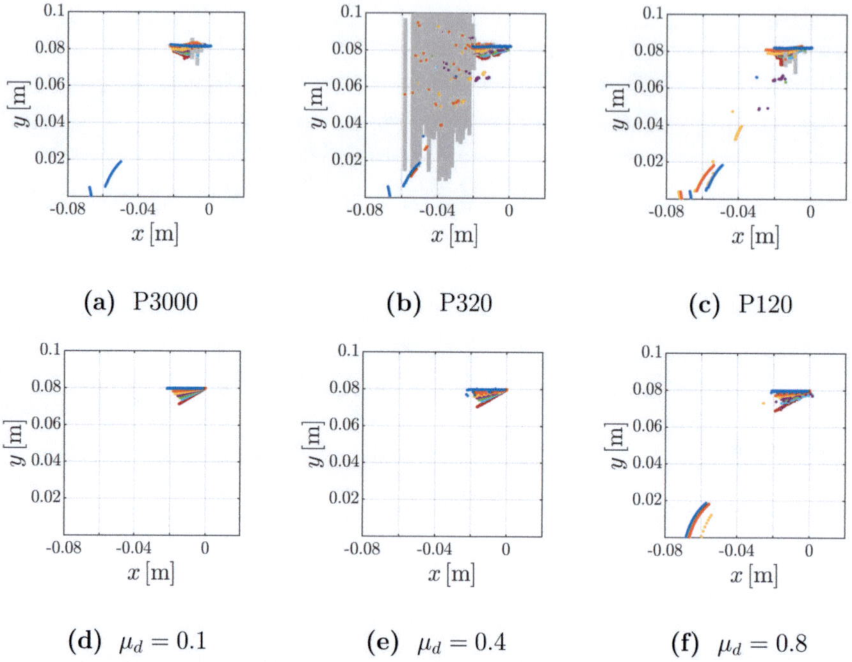

Figure 5.36 – The reconstructed object shapes in **(a)**-**(c)** belongs to the experiments and the ones in **(d)**-**(f)** to the simulations. The reconstructed object shapes are colored in dependence on α: blue- 0°; red- 5°; yellow- 10°; magenta- 15°; green- 20°; cyan- 25°; dark red- 30°. In **(a)**-**(c)**, the grey shade enveloping the reconstructed shapes correspond to the standard deviation of P_y.

Only the values for P320 are smaller than the expected ones. Furthermore, these values show an immense standard deviation, which is already notable for the reconstructed object shapes in Figure 5.36(b). For the α related to retraction, in Figure 5.37, the presence of inflection points along the elasticae cause still large outliers of α. But, also if there are no inflection points, the values of α are poorly determined as before for protraction. Here, the values are too small compared to the expected values. Since the values regarding protraction are predominantly too larger and the ones belonging to retraction too small, there seems to be a systematical influence that is related to the direction of displacement. This effect cannot be observed for the simulation data in Figure 5.37(c) and (d). Here, all values, excepting the ones of large μ_s and small α, match well the expected values.

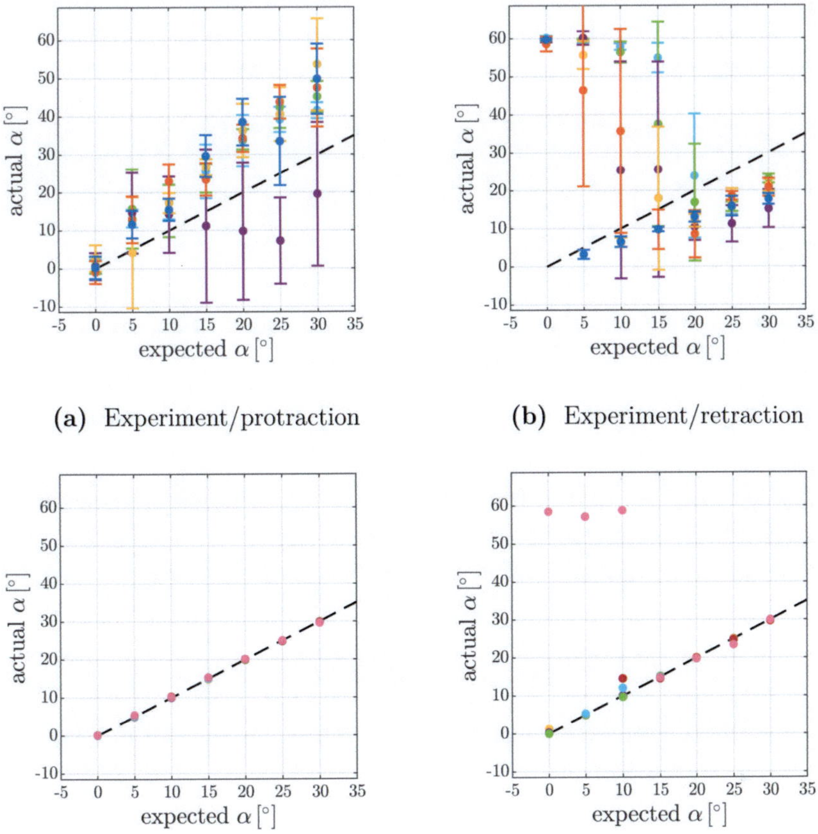

(a) Experiment/protraction

(b) Experiment/retraction

(c) Simulation/protraction

(d) Simulation/retraction

Figure 5.37 – The determined α are compared to the expected values, for different sandpapers in **(a)** and **(b)**: blue- P3000; red- P1000; dark yellow- P600; magenta- P320; green- P180; cyan- P120 and for different μ_s in **(c)** and **(d)**: blue- 0.1; red- 0.2; dark yellow- 0.3; magenta- 0.4; green- 0.5; cyan- 0.6; dark red- 0.7; pink- 0.8. The black, dashed line indicates the expected values.

Determining μ_d applying (A.7) considering the determined values of α, leads to the alike conclusions as before in the case of α. For the experimental data, the detected values of μ_d vary extremely and a reliable detection is not possible, neither in pro- nor retraction. In high contrast to these findings, μ_d is well detected using the simulation data, excluding the values that belong to elasticae exhibiting originally an inflection point.

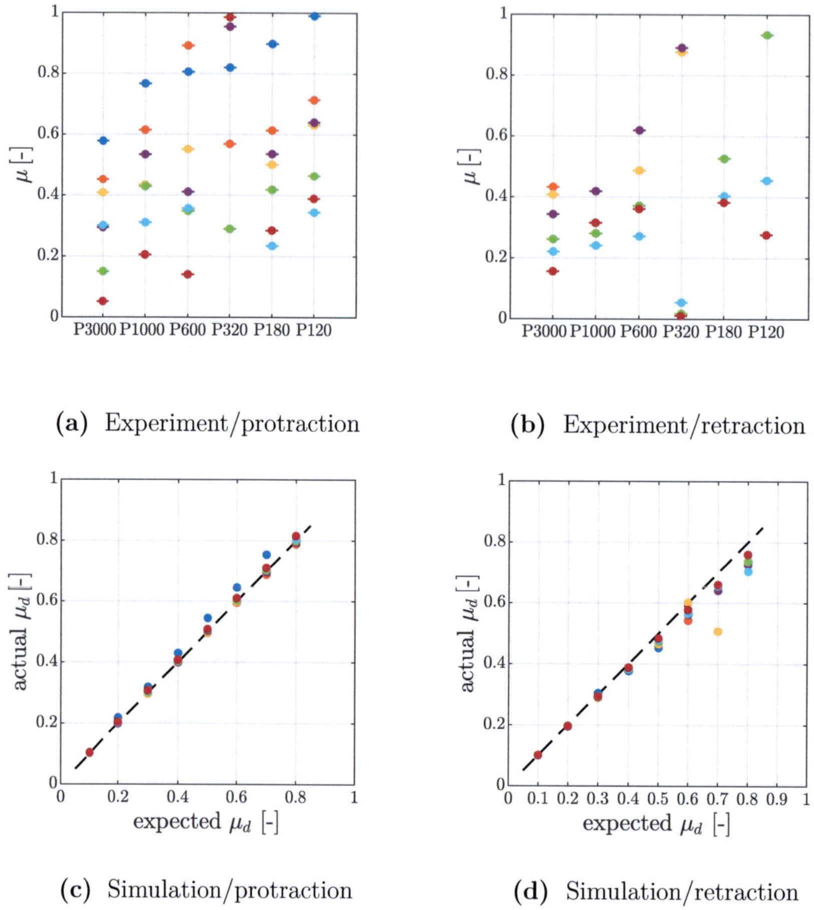

Figure 5.38 – The values of μ_d for all types of MICTs and η are shown with: blue- 0°; red- 5°; dark yellow- 10°; magenta- 15°; green- 20°; cyan- 25°; dark red- 30°. The dashed, black line corresponds to the expected values.

This implies, that there is not a general issue in the used approach to determine μ_d. The experimental setup reaches its performance limit for the present experiment. Finally, also the number of stick-slip events/peaks during a scanning sweep is analyzed, see Figure 5.39. Here, a larger grit size/μ_d a smaller number of peaks can be concluded for all cases. The quantities of detected peaks for the different MICTs are comparable to the ones in Figure 5.33. The number of peaks changes for a change of α which is clear because a change of α always comes up with a change of the actual distance between sensor base and contact point.

Generally, the spread between the values for one MICT is smaller than for the results in Figure 5.39. Summarizing this subsection:

- The influences of a MICT can be analyzed for $\alpha \neq 0°$.
- There is a difference in pro- and retraction of the sensor for support reactions, which is alike to the one in Subsection 5.3.1.
- The validity of the experimental results is strongly limited but the results regarding the simulations demonstrate the applicability of the proposed approach.

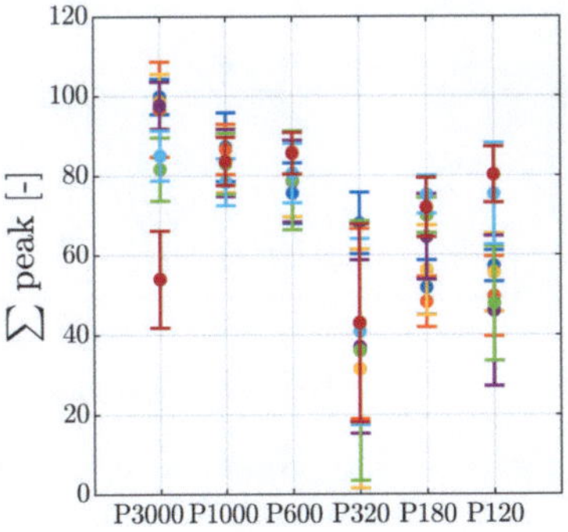

(a) Experiment/protraction

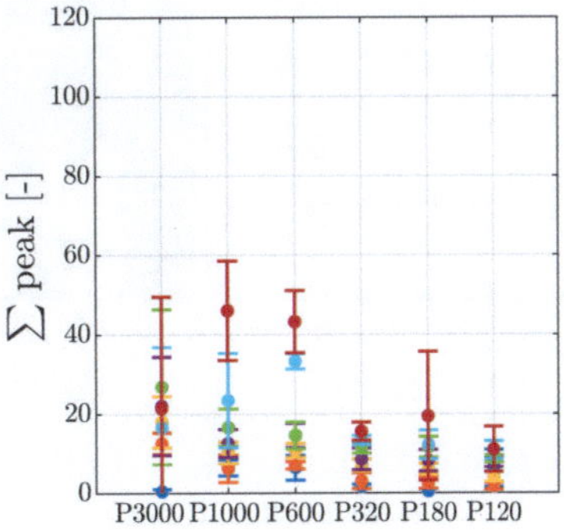

(b) Experiment/retraction

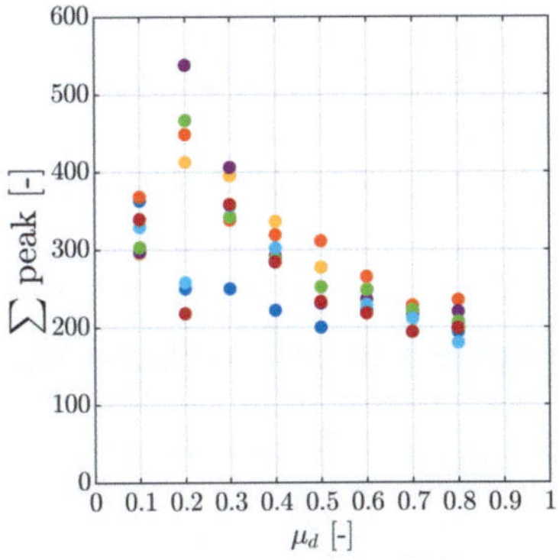

(c) Simulation/protraction

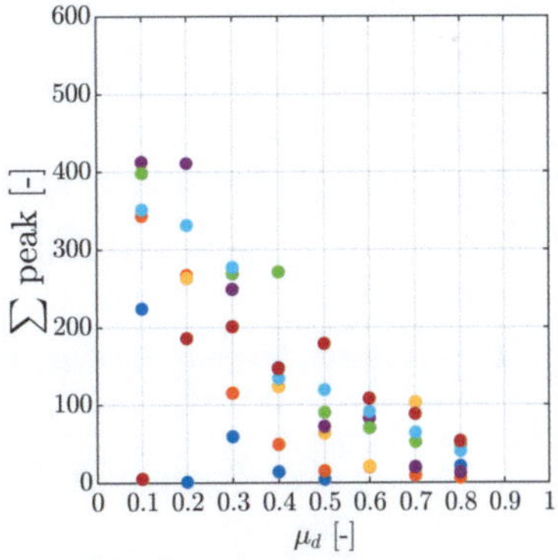

(d) Simulation/retraction

Figure 5.39 – The number of detected peaks of M_z for all types of MICTs and α are shown with: blue- 0.04 m; red- 0.05 m; dark yellow- 0.06 m; magenta- 0.07 m; green- 0.08 m; cyan- 0.09 m.

5.3.3 Sinusoidal shape with microscopic surface texture

In a final step, the test object f_2 is scanned considering a MICT. In the Subsections 5.3.1 and 5.3.2, the object shapes are characterized by a constant α. Here, α is changing along the object shape. The MICTs are represented by a sandpaper of type P320 in experiment and $\mu_d = 0.75$ in simulation.

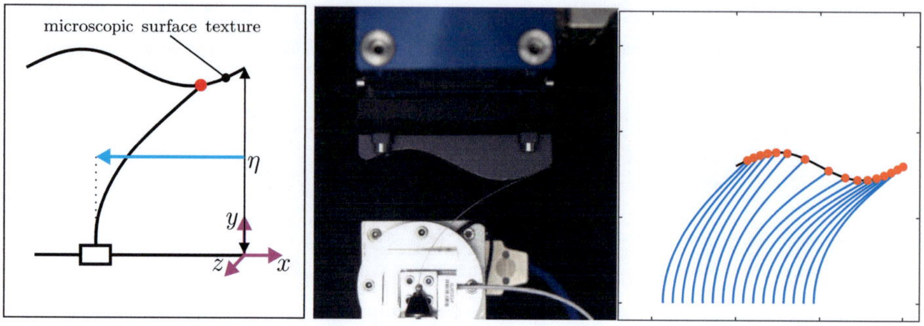

(a) Sketch (b) Experiment (c) Simulation

Figure 5.40 – Scanning the test object f_2: **(a)** sketch of the present scenario; **(b)** experimental setup (P320), protracting the sensor; **(c)** simulation ($\mu_d = 0.75$). In **(a)**, the cyan arrow indicates protraction. In **(c)** the red dots correspond to the current position of the contact point along the sensor shaft (colored in blue) and the black line to the expected object shape.

The value of μ_d is chosen based on the findings in Subsection 5.3.1 in order to maximize the similarity between experiments and simulation. The test object f_2 is scanned for $\eta \in \{0.065\,\mathrm{m}, 0.08\,\mathrm{m}\}$ and sensor is protracted only since a retraction is not beneficial for this type of object shape, see Figure 5.40(a).

Table 5.1 – Determined values of μ_d in experiment (P320) and simulation ($\mu_d = 0.75$).

Microscopic surface texture	η [m]	actual μ_d [–]
P320	0.08	0.76
P320	0.065	0.735
$\mu_d = 0.75$	0.08	0.78
$\mu_d = 0.75$	0.065	0.75

In the Figures 5.40(b) and (c), the elasticae are very alike to the ones in Figure 5.12. Comparing the support reactions in Figure 5.13 to the ones in Figure 5.41, it is obvious that the extreme values of the simulated support reactions strongly increase by the presence of friction. Therefore, the simulated support reactions match the experimental ones very well in Figure 5.41. In experiment and simulation, the low-frequency part of the signals is superimposed by a high-frequency sawtooth-like oscillation.

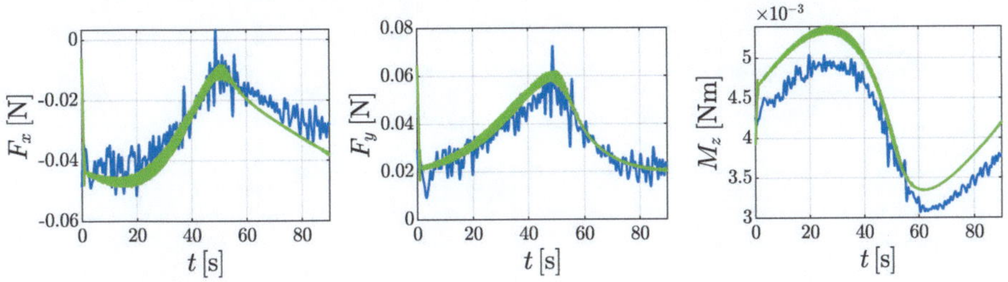

(a) $\eta = 0.08\,[\mathrm{m}]$ **(b)** $\eta = 0.08\,[\mathrm{m}]$ **(c)** $\eta = 0.08\,[\mathrm{m}]$

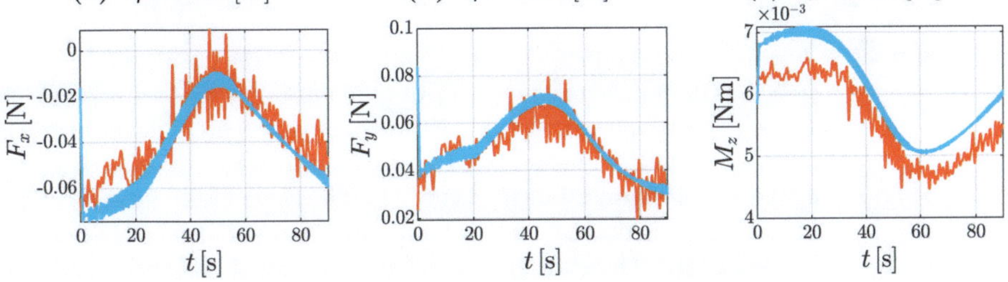

(d) $\eta = 0.065\,[\mathrm{m}]$ **(e)** $\eta = 0.065\,[\mathrm{m}]$ **(f)** $\eta = 0.065\,[\mathrm{m}]$

Figure 5.41 – In **(a)**-**(c)** and **(d)**-**(f)**, the support reactions F_x, F_y and M_z are illustrated for $\eta \in \{0.065\,\mathrm{m}, 0.08\,\mathrm{m}\}$. The blue (experiment) and green (simulation) curves correspond to $\eta = 0.08\,\mathrm{m}$ and the red (experiment) and cyan (simulation) ones to $\eta = 0.065\,\mathrm{m}$.

Based on the findings of the previous subsections, this high-frequency oscillation can be identified as stick-slip effect. Reconstructing the object shape and determining the corresponding α (see Figure 5.42), the values of μ_d are successfully determined, see Table 5.1. In the Figures 5.42(b) and (c), the influence of friction is observable by a approximately constant offset of 40° of the curves in Figures 5.42(c) compared to Figures 5.42(b). The found value corresponds to ζ, see (A.7). The peaks are not further evaluated since there is no reference value to compare with. But, there are notable stick-slip events that can be evaluated. In conclusion:

- For complex object shapes with varying α the present μ_d and/or the number of peaks can be determined.

- The object shape is encoded in the low-frequency part of the support reactions and, therefore, also μ_s is related to the low-frequency signal parts, see Figure A.4, too.

- The high-frequency part of the support reactions is important while evaluating the number of stick-slip events/peaks but not for detecting μ_d.

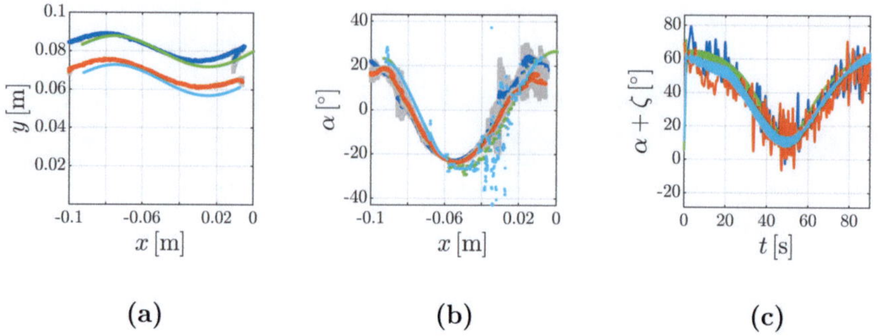

Figure 5.42 – The blue (experiment), green (simulation) curves correspond to $\eta = 0.08\,\text{m}$ and red (experiment), cyan (simulation) ones to $\eta = 0.065\,\text{m}$. In **(a)**, the reconstructed shape of test object f_1 are shown for both levels of η. The grey shade enveloping the reconstructed shapes correspond to the standard deviation of P_y. The slope of the reconstructed object shape is determined by numerical differentiation of the object shape in **(b)**. The sum of α and ζ is determined in **(c)** by applying (A.7) on the support reactions.

5.4 Discussion & interpretation

The main objective of the present work is summarized by the question: „Why is a natural vibrissa such a good sensor?", see Chaper 3. The literature review in Chapter 2 and the systematic characterization of the natural paragon (Section 4.1) highlight that one determining factor regarding this principle question are the inherent characteristics of a natural vibrissa. These inherent characteristics are an essential part of the stimulus leading apparatus [Carl et al.; 2008] and cause a pre-processing of the gathered information [Bagdasarian et al.; 2013].

5.4.1 Object shape recognition

The matter concerning object shape recognition includes two major aspects, which can be strongly merged. On the one hand, there is information about the location of the object and, on the other, information about the object shape. Both information can be qualitative or quantitative.

For a localization task, due to a contact, it can be determined if the object is located at the left or the right side without further knowledge about the distance. Similarly, it can be detected that an object shape is round without knowing the actual diameter of, e.g., the sphere. Which type of this information is necessary depends on the scanning task. In this work, a combination of object localization and object shape recognition is used. The sensor is swept along the object shape for a defined period and the support reactions are recorded with a defined sampling rate and stored. They are converted into a finite number of contact point locations, which represent the scanned object shape by post-processing. During this procedure, first, the information about the spatial position of the object gets converted into elastic energy/support reactions, and, afterward, the support reactions are converted back to the spatial information, compare Figure 4.4. Here, the transitions from a qualitative information to a quantitative one is realized by the sensor shaft and connected sensor periphery. The material and geometric properties of the sensor shaft are the inherent scale of the present sensor system. Without exact knowledge about, e.g., the present length of the sensor shaft or its Young's modulus, the quantitative information is erroneous. In Subsection 5.1.2, for experiments, the qualitative information „the object shape is straight and inclined" is well detected but the actual location of the object differs from the expected one. This holds true, in different magnitudes, for nearly all reconstructed object shapes in regarding experiments the Sections 5.1, 5.2 and 5.3. Therefore, this systematical deviation may be related to differences regarding the inherent scale of the sensor system like a deviation of the actual Young's modulus. Another influence can be the reference coordinate system and guides, even if the test objects are located with high diligence the actual position may be different from the expected one. Furthermore, the precision in positioning (value of x_0) of the guides affects the results. The data belonging to the simulations are not affected by these problems because all mentioned problems are probably caused by imperfections of the experimental setup. In the present work, the reconstruction is most successful when the sensor is protracted along object shape and η is small. If there are MACFs, retracting the sensor is not analyzed because of restricted reasonableness. Expressing η in dependence on the length of the sensor shaft L:

- $\eta \approx 0.9 \cdot L$: for an extensive MICT, the reconstruction fails due to extensive sticking while retraction, but works good for protraction;
- $\eta \approx 0.6 \cdot L$: there are distinct stick-slip oscillations affecting more the retraction than the protraction;
- $\eta \leq 0.457 \cdot L$: the contact between test object and sensor shaft is always a tangential one, the stick-slip oscillations get less distinct as well as the differences between pro- and retraction are small.

The idea to retract the sensor without losing contact with the object seems to be without advantage for this process. Furthermore, a smaller η corresponds to stronger low-frequency parts of the support reactions, which are relevant for the procedure of object shape recognition.

Comparison to natural vibrissae and other alike sensing tasks:
With a view to natural vibrissae, the observation that the low-frequency parts of the support reactions are mandatory for the procedure of shape recognition may be associated with the difference of RA and SA mechanoreceptors which encode distinct information of gathered stimuli [Leiser, Moxon; 2007]. The authors of [Maravall, Diamond; 2015] reported that the captured signal can be divided in a *slow bending component* and a *fast transient component* and, in [Jenks et al.; 2010], it is assumed that rats can simultaneously process low and high-frequency signals during exploration. Moreover, in [Carvell, Simons; 1995], a stronger deflection of the vibrissa hair shaft is related to stronger signals of SA mechanoreceptors, which can be interpreted as a pre-processing of the signals [Bagdasarian et al.; 2013]. This observation becomes further support by the shape of a natural vibrissa because the tapering increases this effect. This a hint to a low/high-pass filtering of the signals as done in the data processing regarding the data used in the present work and that a small η supports the signal components belonging to the object shape. Furthermore, for localization and reconstruction of the object shape, the information about x_0 and the associated reference frame are necessary and, as mentioned, may be crucial for the quality of the results. In the case of vibrissae, these issues are handled by the proprioceptive systems. For example, mechanoreceptors continuously record the position of the vibrissae hair shaft [Szwed et al.; 2003] which allow for a transformation of the signals between whisker centered and head centered coordinates [Hartmann; 2011], [Hartmann; 2015]. In [Hartmann; 2009], performing a specific exploration task is related, i.a., to an open-loop control of the movement of a vibrissa. This means, the scanning trajectory must be planned by the animal before starting the scanning process and it must be ensured that the object is touched in a reasonable way. Therefore, it seems to be possible that:

- first, the object location gets determined by an initial contact as described in, e.g., [Baldeweg et al.; 2014];
- followed by a specific movement along the planned trajectory that depends on the actual scanning task.

For example, as found in this work, if the overall object shape is the information of interest, the planned trajectory should consider a small η. The findings in [Grant et al.; 2008] confirm this idea because rats seem to make use of their macro vibrissae to detect the distance to an object and orient their micro vibrissae for a detailed scanning sweep, see Subsection 2.1.2. Furthermore, the authors of [Brecht et al.; 1997] mentioned that rats use their micro vibrissae to recognize object shapes. This two-step procedure highlights again that the task of localization and object shape recognition are connected, even if the finally generated information may be independent. So, if the desired information is the object shape, it is not necessary to know its location, compare [Diamond et al.; 2008b], [Garrett; 2009]. However, to detect the object shape in a suitable way, it seems to be necessary to know the location of the object for planning a proper scanning trajectory.

This step is partly skipped in the present work because the distance between sensor support and object is always chosen in an appropriate manner. For an autonomous sensing task, this is nearly impossible. The support reactions get recorded for the entire scanning sweep, as by the force and torque sensor used in this work. Consequently, a connection between actor and sensor system (compare Figure 2.7) is necessary to relate captured support reactions to a reference coordinate frame. In [Fassihi et al.; 2015], [Bush et al.; 2016] and [Fassihi et al.; 2017], it is reported that rats can integrate the record signals over time, that they combine actor and sensor signals and store the recorded signals.

There are findings in literature that are in slight contrast to the discussed procedure of object shape recognition. In [Lederman, Klatzky; 1987], it is found that the tactile exploration of an overall object shape with fingers is done by following its contour. An alike approach is used in industry in form measurement tasks [VDI/VDE 2617]. Here, a stiff probe is displaced considering a constant probing force along the object shape, which results in a contour following scanning trajectory. In context to natural vibrissa or an artificial vibrissa-like sensor, similar observations are done [Horev et al.; 2011]. The natural vibrissal system includes control-loops combining sensor and actor information [Kleinfeld et al.; 2006]. A closed-loop control is discussed in [Hartmann; 2009], which allows for real-time sensor-actor feedback. Therefore, tracking the support reactions, a scanning trajectory following the object shape can be realized (perceptual constancy) [Saraf-Sinik et al.; 2015]. This trajectory may be recorded by the proprioceptive system and contains information about the object shape, but not necessarily about the object location. To recover this information, in [Solomon, Hartmann; 2011] multiple alternatives to the algorithm of [Scholz, Rahn; 2004] (see Section A.1) are presented. One of those uses the first time derivative of M_z to determine the distance to the object. The first time derivative of forces and moments, the so-called yank, is a frequently used variable in the sensor-actor system in biomechanics [Lin et al.; 2019]. In engineering, this value is known as jerk [Dresig, Holzweißig; 2009]. This could be a hint for different mechanisms regarding object shape detection. Compared to the procedure used in this work, this approach requires more effort in actuation but less effort in post-processing. The variety in whisking and scanning behaviors reported in Subsection 2.1.2 do not approve this discussion but also do not refute it.

5.4.2 Detection of macroscopic surface textures

The limits between overall object shape, MACFs and MICT are not exactly defined. Therefore, MACFs can be encoded in both the low- and high-frequency part of the support reactions depending on the used definition. These limits have to be interpreted in dependence on the geometric properties of the sensor shaft. The findings in Section 5.2 demonstrate that MACFs can be detected using the time-series signals of the support reactions, as well as by analyzing the reconstructed object shapes. The sought information of the MACFs decides which of the two methods is most appropriate.

As in context of the object shape recognition, η has a mandatory influence on scanning MACFs that superimposes a straight horizontal object, see Subsection 5.2.1. Considering a scanning trajectory tangential to the overall object shape and an orientation of the sensor shaft normal the overall object shape, the following statements are derived:

- $\eta \approx 0.9 \cdot L$: the sensor tip contacts the MACFs, but the amplitudes of the support reactions are very low;
- $\eta \approx 0.8 \cdot L$: the sensor tip contacts the MACFs and the influences of the MACFs are very well recognizable in the support reactions;
- $\eta \approx 0.6 \cdot L$: there are multiple contacts between sensor shaft and the MACFs, and reliable scanning results are difficult to obtain;
- $\eta \leq 0.457 \cdot L$: the contact between the MACFs and the sensor shaft is always tangentially one, so, e.g., small grooves are not contacted by the sensor and cannot be reconstructed.

Consequently, a successful determination of the properties of the MACFs requires a control of the distance between sensor shaft's support and object. If this distance is too small, the bending induced curvature of the sensor shaft works like a morphological filter (mechanical low-pass filter), and the MACFs cannot be detected. Simultaneously, the low-frequency parts of the support reactions are increased by the small distance, and, therefore, the signal parts containing the information regarding the overall object shape get prominent. In addition, the orientation of the sensor shaft must be controlled in the way that the contact takes place at or close to the tip of the sensor shaft. To maintain a constant influence of this morphological filter, it is necessary that the scanning trajectory follows the overall object shape, and that the relative orientation between overall object shape and sensor does not change. If these conditions are not fulfilled, the influence changes during the scanning process, and the information about the MACFs becomes adulterated.

Comparison to natural vibrissae and other alike sensing tasks:

Reports regarding the natural paragon confirm these findings. Generally, rats can detect MACFs as demonstrated in, e.g., [Zuo et al.; 2011]. Following [Carvell, Simons; 1995], it is more likely that the MACFs are encoded in a spatial signal than in a time-series signal. This finding matches the results of Section 5.2 because not all information regarding the MACFs is contained in the support reactions. To provide all information the reconstruction algorithm must be applied and to convert the support reactions into spatial information again. Furthermore, rats seem to optimize/control the *amount of bending* regarding their vibrissae hair shafts while touching an object [Carvell, Simons; 1995]. The hypothesis of a *minimal impingement* in [Mitchinson et al.; 2007] confirms these findings.

Both works are further hints for the use a closed-loop control, which seems to be necessary to guarantee suitable scanning conditions and reliable results like already mentioned in Subsection 5.4.1. Considering the geometrical shape of a natural vibrissa, the tapered shape hampers the detection of MACFs, if the object is located at a larger distance because the signal amplitudes are very low. This is in accordance with the findings in Section 5.2. Besides the influence of the tapered shape, also the inherent curvature affects the detection of MACFs. As well as the bending induced curvature, the inherent curvature serves as a mechanical low-pass filter with the difference that it is not adaptive to the distance to the object. Weber's law is important for most natural sensing system and, therefore, also for the vibrissal system [Solomon, Hartmann; 2011, pp. 3054]: „Weber's law holds that the resolution or 'just noticeable difference' (JND) of a stimulus intensity is proportional to the stimulus intensity". In short, if a stimulus is already strong, a fine change cannot be detected well. With a view to the present findings, this implies that when η is small fine MACFs cannot be recognized well because the low-frequency signal parts that belong to the overall object shape are large. Because of its physiological origin, this effect cannot be confirmed by the present work, but it is further evidence that the distance between object and vibrissa base/sensor support seems to be an actively controlled parameter. Hence, the elastic deformation of the vibrissa hair/sensor shaft is an inherent characteristic of the sensor system and part of the pre-processing of the captured signals.

5.4.3 Detection of microscopic surfaces textures

In engineering and surface metrology, the profile method is frequently used to analyze a MICT, see Subsection 2.2.2. There, the tip of the scanning probe is displaced along the object and the contour of the MICT is determined, whereby the information about the overall shape is filtered out. Out of the perspective of the present work, the profile method is more like a shape reconstruction or MACF detection task on a small scale. Here, the tip of the sensor shaft works also like a morphological filter because, e.g., a groove can only be detected if its width is larger than the sensor shaft diameter. So, the tip of the sensor shaft already filters the collected information before the curvature starts contributing to a pre-processing of the signals. This influence of the tip is not adaptive, its diameter cannot change and can be seen as a fixed limit of resolution between MACFs and a MICT. Therefore, the sandpapers used in experiments are considered as a MICT. Thus, a further inherent characteristic of the sensor that contributes to a pre-processing of the information is the tip diameter.

In this work, the MICT is analyzed in terms of the present μ_d and number of stick-slip events/peaks for a fixed scanning distance. Both parameters are closely related due to the theory of dry friction, see Subsection 2.2.2. In the case there are MACFs and a MICT (not analyzed in the present work), the concerning signals can be distinguished by their waveform because the distinct shape of the sawtooth-like trend of the signals belonging to the MICT, which is different to the spike-like waveform found for MACFs.

Out of the support reactions, an estimated μ_d can only be determined directly when the used scanning trajectory is tangential to the scanned surface, see Subsection 5.3.1. If not, first, the scanned object shape and its slope must be reconstructed to determine μ_d. Here, the low-frequency parts of the support reactions are used, compare Figure A.4. The superimposing stick-slip oscillation contains information about μ_s, which can be determined by analyzing the extrema of the stick-slip events (not included in the present work). In Figures 5.32 and 5.38, for the simulations, systematic deviations can be observed for large μ_d. Since the values of μ_d tend to be too large for protraction and too small for retraction, an averaging method may be necessary to exclude the influence of the systematic deviations. Such an approach is comparable to the well-known reversal method used in engineering. For experiments, there seem to be further effects like an interaction between the sensor shaft tip and contact surface, which is not included within the limits of dry friction. For a successful detection of μ_d, it is necessary that the sensor shaft moves across the surface, when the tip only sticks a detection is not possible because inflection points along the sensor shaft can occur [Steigenberger et al.; 2015].
The number of peaks is contained in the signals of the recorded support reactions in form of a sawtooth-like oscillation superimposing the low-frequency signal part as already found in [Hipp et al.; 2006]. The first time derivative of M_z (numerical differentiation) is used to detect the stick-slip events. Due to the abrupt changes in the time-series signal, the stick-slip events become easily detectable by large peaks. For low friction coefficients between sensor shaft and MICT, there are many peaks, with an increasing friction coefficient the number of peaks decreases, see Figures 5.33 and 5.39. Moreover, the combination of η and pro-/retraction seems to influence the number of peaks as well. Consequently, the number of peaks alone is not a reliable parameter to describe a MICT, it has always be combined with information about η or α. Furthermore, also v may influence the number of peaks but this is not investigated in the present work. In contrast to the analysis of μ_d, the occurrence of inflection points is less important for the detection of the number of peaks. If the tip only sticks, the number of peaks is zero, indicating extensive sticking, but μ_d will be any value. Therefore, the number of peaks is robust against this effect. The number of peaks depends on the conditions inside the contact between sensor shaft and object and not only on the MICT.
Both analyzed parameters show that there is a difference between pro- and retraction. In the case of μ_d, this observation is not advantageous. In contrast, for the number of peaks, the peaks get more distinct and can be detected better. Furthermore, also contact pairs including low coefficients of friction cause detectable signals [Merker et al.; 2018]. But, retracting the sensor is less stable than protracting it because of effects like the turn-over [Scharff et al.; 2017a].

Since sandpapers are scanned, it must be mentioned that the influence of wear of the sensor shaft and, also, of the MICT is totally neglected in the present work.

Supposing that a horizontal straight object with MICT is scanned using a scanning trajectory tangential to contact surface and an orientation of the sensor shaft normal to the contact surface, the influences of η and the direction of displacement can be summarized:

- $\eta \approx 0.9 \cdot L$, protraction: the amplitudes of the support reactions are very low, the frequency of the stick-slip oscillation is high, and the peaks are difficult to detect;

- $\eta \approx 0.9 \cdot L$, retraction: the sensor tip sticks on the MICT, the support reactions get large and there are no stick-slip oscillations, this is easy to detect but lacks information;

- $\eta \approx 0.6 \cdot L$, protraction: there is a uniform stick-slip oscillation with a higher frequency and low amplitudes, which is detectable well;

- $\eta \approx 0.6 \cdot L$, retraction: there is a uniform stick-slip oscillation with a lower frequency and higher amplitude, which is detectable very well;

- $\eta \leq 0.457 \cdot L$, equally, for pro- and retraction: the frequencies of the stick-slip oscillations get very high and the amplitudes very low, the peaks are difficult to detect.

Considering (2.1), it becomes clear that there is no direct relation between η and the contact conditions. Due to the bending stiffness of the sensor shaft, F_n scales with η and, consequently, the contact conditions get affected. Therefore, best conditions to scan a MICT with the present sensor concept include a horizontal surface with $\eta \approx 0.06\,\mathrm{m}$, an orientation of the sensor shaft normal to the contact plane, and a retraction of the sensor while the scanning trajectory is tangential to the scanned surface. Since the recorded signals can be manipulated by the displacement direction of the sensor, this is another inherent characteristic of the sensor system because this can be used as, e.g., signal amplification when the number of peaks is used as classifier. But, the only useful application of a retraction is when a flat horizontal surface is scanned, in all other cases, investigated in this work, a protraction leads to more reliable signals. Furthermore, to guarantee a comparability between the detected number peaks of different MICTs, it is necessary that they are scanned using an equal scanning velocity, orientation of the sensor, displacement range/scanning time, and distance between sensor support and object. If the orientation of the sensor shaft is always normal to the contacted surface, μ_d can always be determined as well as the number of peaks. Especially, if a MICT is scanned by pro- and retracting the sensor, there are two different sets of signals for the same MICT. Consequently, to encode the MICT using only the support reactions becomes impossible. Nevertheless, if these conditions are met, the number of peaks is the better classifier considering the results of Section 5.3 because the experimental results for μ_d are weaker.

Comparison to natural vibrissae and other alike sensing tasks:
The results of the present work share several similarities with observations regarding natural vibrissae. The diameter of a tip of a natural vibrissa is normally smaller than the one of the sensor shaft used in this work and is related to the detection of small texture features [Williams, Kramer; 2010]. But, its role as a mechanical low-pass filter/morphological filter does not change. The present findings regarding the number of peaks confirm the ideas of *kinetic signature hypothesis* and *Slip hypothesis*, see Subsection 2.1.3. The most evident points are:

- rats seem to control the velocity of the touching vibrissae while scanning a surface [Carvell, Simons; 1990];
- the information regarding the actuation must be considered while active exploration [Diamond, Arabzadeh; 2013];
- the authors of [Hipp et al.; 2006] demonstrated that the low-frequency part of the recorded signals are superimposed by higher-frequency stick-slip oscillations;
- the signals can be divided in low-frequency and high-frequency parts [Leiser, Moxon; 2007], [Maravall, Diamond; 2015];
- the asymmetry between the signals of pro- and retraction is observed in [Fend et al.; 2003] and [Carvell, Simons; 1990];
- sandpapers are recognized based on the number of peaks in, e.g., [Wolfe et al.; 2008], [Oladazimi et al.; 2018].

The mentioned six items are consistent with the results of the present work, which confirms that several of the inherent characteristics of a natural vibrissa are included in the present design of the artificial sensor. Including more details of, e.g., the geometric shape of a natural vibrissa probably influences some of the mentioned effects but will not generally change them. For example, including a tapered shape influences the stick-slip events [Williams, Kramer; 2010], or considering the inherent curvature can cause further effects as parametric resonance [Volkova et al.; 2016]. Another hint must be highlighted, in the present work the first time derivative of M_z is determined by numerical differentiation (see Section 5.3), which is similar to the yank in [Lin et al.; 2019]. But in, e.g., [Campagner et al.; 2018], it is reported that the mechanoreceptors probably encode the support reactions. Out of the perspective of mechanics, this makes sense because the base of the vibrissa hair shaft deforms the surrounding elastic tissue when the hair shaft gets loaded by a contact force. Therefore, the mechanoreceptors start recording since they get deformed, too. Hence, a further mechanism must provide the determination of the first time derivatives of the support reactions. This can probably be done by the connected neuronal periphery. The conclusion that knowing only the number of peaks is not enough to detect a surface texture is confirmed in [Wolfe et al.; 2008] and [Wiertlewski; 2013].

In the present work, this issue is solved since the distance between sensor support and surface and the orientation of the sensor shaft are known a priori, as well as, that the displacement velocity is equal and constant for all scanned surfaces. But, for an unknown object and environment, this information is not necessarily available. Therefore, the vibrissal system/artificial sensor system must provide mechanisms to fulfill these conditions. One way is to combine the information gathered while scanning the same object various times. Rats can use information captured during multiple touches [Sofroniew et al.; 2015] and, consequently, a step-by-step analysis of the unknown object can be assumed like:

- zeroth - localization of the object like in, e.g., [Baldeweg et al.; 2014];
- first scan - shape recognition;
- second scan - MACFs;
- third scan - MICT.

Following this scheme, the scanning trajectories can be planned in order to fulfill the requirements of the aimed task within the concept of an open-loop control [Hartmann; 2009]. This fits in the discussion regarding the recognition of an object shape in Subsection 5.4.1.
In [Lederman, Klatzky; 1987] the tactile interaction of a fingertip with a surface is related to a lateral motion. In the present context, a lateral motion is a motion tangentially to the overall object shape. As discussed before in Subsection 5.4.1, an approach considering a real-time closed-loop control is may be beneficial to detect properties of a MICT as well [Hartmann; 2009], [Saraf-Sinik et al.; 2015]. To realize a scanning trajectory that is always tangential to the overall object shape the support reactions can be tracked and adaptive. Thus, the scanning trajectory fulfills the requirements found to be necessary for a successful determination of properties of the MICT. An alike procedure is used in surface metrology, there, e.g., for the profile method, a constant probing force is specified [ISO 3274; 1996].
One more possibility is that rats sweep their vibrissae only for short periods along the object and, consequently, the changes of the contact conditions are low and the information about a MICT can be integrated over several sweeps [Diamond, Arabzadeh; 2013]. However, this approach seems to be less accurate because the object shape in combination with the scanning time will strongly influence the results.

Even if the discussed conditions to scan a MICT are respected, there is a further issue. If two MICTs are scanned with respect to identical conditions, it can be detected if the first one is different from the second one, but there is a lack of comparability to a more general classification. This problem is illustrated in Figure 5.43. To do a classification a reference is necessary. In surface metrology, sensors are calibrated using surface etalons.

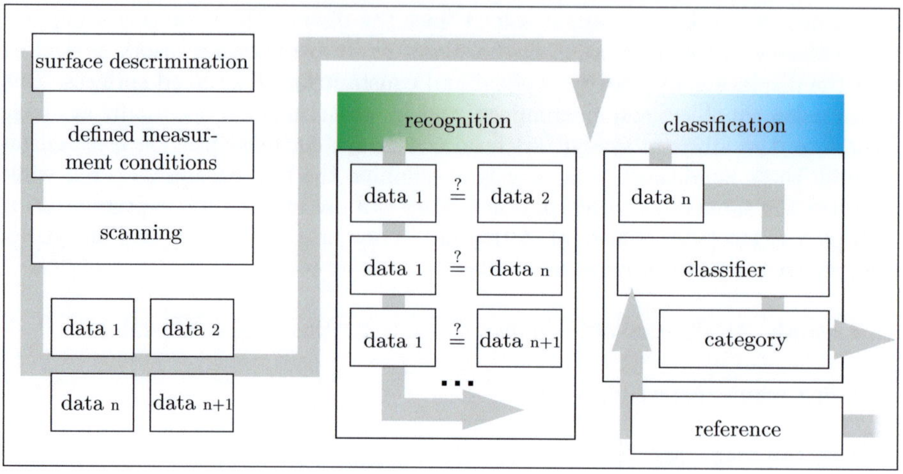

Figure 5.43 – Scheme to distinguish between recognition and classification task in surface texture detection following [Cutkosky et al.; 2008].

All following measurements are done in comparison to this calibration. However, a natural vibrissa is probably not calibrated using any surface etalon. This problem is alike to the discussion regarding the transition from qualitative to quantitative information for the recognition of the shape of an object. There, the geometric and material properties of the sensor shaft serve as a scale. For the present case, the actor signals during free whisking may serve as a reference, and/or stored memories of former scans are used [Diamond, Arabzadeh; 2013]. The results of this work do not treat this problem. Here, the MICTs are only recognized and not classified.

6 Conclusions & outlook

The present work is done within the scope of the research group regarding tactile sensing with vibrissa-like tactile sensor at Technische Universität Ilmenau and at Pontifical Catholic University of Peru. In a retrospective view, the dissertation [Carl; 2009] focuses on natural vibrissae and their properties, the habilitation [Behn; 2013] extended these investigations to basic concepts regarding mathematical and mechanical modeling of tactile hairs resulting in a project funded by the Deutsche Forschungsgemeinschaft (DFG): „Technische, nicht-visuelle Charakterisierung von Substratkontakten nach dem biologischen Vorbild carpaler Vibrissen". Later on, the author of [Will; 2018] optimized the mathematical modeling and analyzed the corresponding mechanical models in detail. This detailed analysis answered several open questions regarding the procedure of the reconstruction of an object shape with an artificial vibrissa-like sensor, but, simultaneously, it revealed the limits of the used approach. Therefore, the present work returned to a closer analysis of the natural example in order to overcome problems and limits identified by the mentioned works with the goal to enhance the understanding of the functionality and characteristics of a vibrissa-like tactile sensor. In Section 6.1, the findings of the present work are summarized and, in Section 6.2, future steps and application examples are discussed.

6.1 Conclusions

The author of [Steigenberger; 2013] discusses two concepts for an artificial vibrissa-like tactile sensor considering an elastic sensor shaft and a rigid sensor shaft that is elastically supported. It is demonstrated that both concepts work well in context to the reconstruction of an object shape, but the procedure regarding the elastic sensor shaft becomes much more difficult compared to the rigid one. Also, conventional tactile sensor uses in most cases elastically supported probes. So, why a natural vibrissae hair shaft is elastic?

These kinds of, still remaining open, questions lead to the goal of the present work. The inherent characteristics of a vibrissa-like tactile sensor are analyzed in order to evaluate their influence on different scanning tasks like object shape recognition and surface texture determination.

The starting point of the present work was a detailed literature review whereby natural vibrissae and tactile sensing were focused. The collected information is systematically evaluated and summarized in the concept of biomechatronic system, see Section 2.2. The biomechatronic system is a concept to describe the natural paragon in terms of a general framework out of the perspective of engineering. Unsurprisingly, it is found that a vibrissa hair shaft and the connected follicle-sinus complex are two very important components of the vibrissal system. Here, the novelty is the twofold role of these components. Both are strongly incorporated in the actor and sensor subsystems. These general findings are further specified using the idea of a stimulus leading apparatus. Following the idea of the stimulus leading apparatus, every component sitting in between the stimuli and neuronal periphery contributes in a particular manner to an extraordinary pre-processing of the captured signals. Categorizing a natural vibrissa out of this perspective led to the design of the artificial vibrissa-like tactile sensor used in this work. Besides the sensor, also the test objects are designed in a methodical way. In this work, the object properties are given by the overall shape, macroscopic surface features, and microscopic surface texture. These three types of information superimpose each other and must be decomposed by the sensor. To do this, 14 different test objects are designed including different overall object shapes, macroscopic surface features, and microscopic surface textures.

The test objects were scanned in experiment with the designed sensor. The experiments were mirrored, as far as possible, by simulations to identify issues of the used data processing and to enhance the understanding of the found effects from the perspective of mechanics in order to facilitate a proper interpretation. To investigate the effects regarding the overall object shape, objects with different shapes and located in varying distances were scanned. It is confirmed that the object shapes are encoded in the low-frequency parts of the support reactions. Using these signals, the object shapes can be reconstructed by knowing the support reactions only. Furthermore, the support reactions scale with the distance between sensor support and object, a closer distance leads to larger support reactions. Consequently, a closer scanning distance amplifies the parts of the signal concerning the overall object shape. A distance of less than 45% of the length of the sensor shaft is found to be most advantageous when a horizontal orientated, flat object is scanned by displacing the sensor along a trajectory that is tangential to the contact plane. The signals regarding the overall object shape get amplified and the signal components belonging to other object properties are quite excluded. Moreover, a retraction without detaching the sensor shaft from the object does not enhance the results. In contrast, it can lead to wrong determined object shapes.

Scanning macroscopic surface features lead to signal components with higher frequencies than the ones belonging to the overall object shape. The macroscopic surface features are analyzed in terms of their number, distribution, and shape. Objects with a straight overall shape superimposed by macroscopic surface features are scanned. Combining a larger scanning distance of approximately 80% of the sensor shaft, with a protraction following a trajectory that is tangential to the overall object shape and a normal to the overall object shape orientated sensor shaft, is most effective to detect properties of the macroscopic surface features. For larger distances, the values of the support reactions are small and, consequently, properties of the macroscopic surface features are difficult to determine. For smaller distances, the bending induced curvature of the sensor shaft hampers a contact between the macroscopic surface features and the sensor shaft. Therefore, the sensor contacts the macroscopic surface features for small intervals or contacts only to the tops of the features and slides over them. The second case occurred also when the sensor shaft was not normal to the overall object shape. Here, also for large distances, the sensor shaft slides over the tops of the macroscopic surface features. It turned out that the diameter of the tip of the sensor shaft defines the limit between macroscopic surface features and microscopic surface texture. For features smaller than the tip, the tip diameter is a morphological filter. However, changes of the microscopic surface texture led to changes of the conditions for the contact between sensor and object. Horizontal straight objects exhibiting different levels of distinct microscopic surface textures are scanned. In experiments, sandpapers with different grit sizes represent the microscopic surface textures. Displacing the sensor along a trajectory, tangentially to the overall object shape and considering that the sensor is oriented normally to the scanned surface, cause stick-slip oscillations superimposing the low-frequency part of the support reactions. The different microscopic surface textures were recognized and distinguished using the present dynamic friction coefficient or number of stick-slip events/peaks as classifier. The number of peaks is the more robust classifier. However, this procedure was only successful when all scans were done under identical scanning conditions, e.g., the distance between sensor support and object. Changes in these conditions caused changes in the results preventing a successful recognition. Also here, the distance between the sensor shaft and the surface is an important parameter. Larger distances and a retraction of the sensor cause inflection points, which can be problematic. For close distances, e.g., smaller than 45% of the length of the sensor shaft, the frequencies of the stick-slip oscillation become very high and the amplitudes small making a successful evaluation of the recorded signals difficult. This holds true for pro- and retraction. For a scanning distance of 60% of the sensor shaft, the stick-slip oscillations were well recognizable. Furthermore, the frequencies and amplitudes changed between the signals of pro- and retraction. A protraction led to signals with smaller amplitudes, higher frequencies, and a retraction to larger amplitudes, smaller frequencies. This effect can be used as signal amplification for the discrimination of fine surfaces and, therefore, it is a further inherent characteristic of a vibrissa-like tactile sensor.

Generally, across all done investigations, it can be concluded that evaluating the support reactions in combination with the actuation parameters is sufficient to determine essential information about an unknown object. Moreover, all discussed inherent characteristics of the sensor shaft are related to its elasticity and deformation behavior. It was possible to observe these inherent characteristics because the actuation and orientation of the sensor were always controlled. Therefore, the inherent characteristics and the related pre-processing of the captured signals are not independent. In contrast, they are strongly connected.

The author of the present work sets up the following hypothesis concerning scanning an unknown object with an artificial vibrissa-like tactile sensor:

- The sensor shaft gets rotated against the object. The signals caused by the collision are used to determine the location of the object.
- In a preliminary sweep, the sensor shaft is protracted using a straight scanning trajectory that considers a distance of 75% of the length of the sensor shaft to the determined object location until it snaps off the object. Then the sensor is retracted to its initial position. The recorded support reactions are used for a first reconstruction of the object shape and to plan the scanning trajectories for the next task.
- Now, the sensor shaft is protracted using a scanning trajectory following the determined overall object shape and considering a distance offset of 45% of the length of the sensor shaft until it snaps off the object. Furthermore, an adjustment of the orientation of the sensor shaft in the way that it is always normal to the contact plane is considered simultaneously to the protraction. After the detachment, the sensor is retracted to its initial position. The recorded support reactions are used for a more accurate reconstruction of the object shape and to plan the scanning trajectory following the overall object shape.
- Next, the sensor support is protracted along a scanning trajectory that follows the well reconstructed object shape with a constant distance offset of 80% and maintaining the normal orientation of the sensor shaft in every moment until the sensor shaft detaches from the object. Afterward, it gets retracted. Based on the captured support reactions the macroscopic surface features can be analyzed.
- In a last step, the sensor is protracted along the same trajectory and with same orientation as before but, this time, with a distance offset of 60% of the sensor shaft length. If the object is straight, the sensor gets retracted without losing contact with the object. If not, it gets protracted until losing contact and then retracted. The measured signals contain information about the microscopic surface texture.

This procedure makes use of the inherent characteristics of an artificial vibrissa-like sensor by controlling the actuation. It demonstrates how an unknown object can be scanned and different properties can be extracted using the same sensor.

Furthermore, there are some remarks drawn from experience regarding the present work:

- The ANSYS simulation should be replaced by a more specific simulation framework because the complexity and restricted information of the underlying algorithms may cause problems, although the used model includes dynamic effects and friction and is usable for a 3D-analysis.
- Neglecting friction in simulation is a very crucial point and will lead in many cases to unrealistic results without much significance for a real sensor.
- Analyzing properties of the natural paragon for only one specific task, e.g., surface texture detection, can cause misinterpretation of the analyzed property because all properties contribute to the multi-functionality of a vibrissa in an inclusive manner. Therefore, investigations reading one particular task and analyzing the influences of, e.g., the inherent curvature must be considered carefully.

Finally, coming back to the initially mentioned example regarding the elasticity of the sensor shaft. It can be answered that the elasticity allows for very different scanning tasks with the same sensor due to its inherent adaptivity. In contrast, a system with a rigid probe is always designed to solve a specific task. Under laboratory conditions, a vibrissa-like sensor may be outperformed by a sensing system with a rigid probe but for, e.g., autonomous robots a vibrissa-like tactile sensor is beneficial due to its multi-functionality.

6.2 Outlook

The present work can be advanced regarding different aspects whereby some points are coinciding. Thus, on the one hand, the collaboration with biologists, neurobiologists, and alike specialists can be enhanced under the aspect to increase the understanding of the vibrissal system of, e.g., rodents. And on the other hand, the transition from a theoretical concept to the design and manufacturing of an artificial vibrissa-like sensor in a technical application can be focused, see Section 6.3. The following remarks can be beneficial and a meaningful next step under both perspectives.

Obviously, the assumptions, restrictions, and constraints in the present work must be dissolved. Hence, a tapered and/or inherent curved sensor shaft must be analyzed considering the findings of the present work. So, the influences of the found effects may change in strength or even in its overall characteristics.

Closely related to the influences of the geometrical shape of a natural vibrissa is its internal structure and surface. There is still very little knowledge about the layered structure and the interaction of the different layers. The findings in [Hartmann et al.; 2003] show that a vibrissa hair shaft is significantly damped. This damping is maybe caused by the interaction between the different layers. In [Popp et al.; 2016] the frictional interaction of layered beam structures is discussed, and it is reported that friction between the layers can lead to high damping ratios. So, a deeper analysis of the layered structure of a vibrissa hair shaft will enhance the scanning performance because the damping is alike to a low-pass filtering and the changes in elasticity can further improve the effects reported in Chapter 5. A very challenging point is the fabrication of an artificial sensor shaft that includes more details of the natural example than the one in the present work.

Also focusing on the scale structure that covers a vibrissa hair shaft can enhance multiple aspects. For example, the influence of the scales on the friction coefficient is anisotropic. This effect is generously neglected in the present work. However, as already discussed in [Voges et al.; 2012], the structure of the scales can contribute significantly to the scanning process. As reported in literature and by the findings of the present work, a vibrissa or a vibrissa-like artificial sensor exhibits an anisotropic scanning behavior. This can be further pronounced considering the influence of the scales. Furthermore, patterns similar to the scale structure of a vibrissa can be found in different conventional products like climbing skins for touring skis, which may be optimized in this way.

Another blank spot in literature is the follicle-sinus complex. The function and influence of its sophisticated structure remain mainly not understood. Out of the perspective of biology, the so-called mechanotransduction is an important point to keep ongoing. A further point is the impact of the visco-elasticity of a follicle-sinus complex on the measured signal. Especially, the function of the blood sinus seems to be interesting in order to control, e.g., the sensitivity of the receptors or in its function as an insulator against external deviations like vibrations. These points become important for an artificial sensor as well. The detection of small forces and torques in 3D-space is still a current field of research. Recent sensor designs allow for a precise, simultaneous measurement of forces and torques, e.g., [Marangoni et al.; 2017]. But their usage as artificial follicle-sinus complex is difficult because of their size and other limitations. In the scope of microelectromechanical systems, there are various sensor designs combining small geometrical dimensions with high sensitivities, e.g., [Rangelow et al.; 2017]. These systems are often restricted to detect signals in only one dimension due to constraints in manufacturing. The authors of [Vaziri et al.; 2007] go as far as they mention the usage of nanotubes as elastic supports at the base of a miniaturized sensor shaft in a way that they work like a torsional spring to determine the current torque. This approach matches well current trends of multiple research fields.

Another approach to adapt an follicle-sinus complex into a technical sensor design can be the use of artificial mechanoreceptors as already demonstrating for artificial skin [Amoli et al.; 2019]. So, the follicle-sinus complex satisfies various requirements that are still challenging to fulfill by sensors available nowadays, and therefore, a closer analysis of the follicle-sinus complex seems beneficial.

Inside the follicle-sinus complex, the transition from mechanical stimulus to action potential takes place, see Figure 4.4. Also, the consideration of neurobiological aspects can contribute in an important way since the signal post-processing strongly influences the final result of a measurement. The surface identification and classification can be improved by adapting parts of the neuronal architecture of rodents as shown in various works, e.g., [Mitchinson et al.; 2007]. Consequently, the sensor performance can be further increased. A modification or replacement of the reconstruction algorithm in order to cover multi-point contacts [Merker et al.; 2020] and inflection points can further contribute to the sensor performance.

Besides a deeper analysis of the structural components and their function, also the whisking/actuation must be further investigated. This includes the uncertainties regarding macro and micro vibrissae as well because the findings in literature are not clear. For both types of vibrissae, there are observations that animals use macro as well as micro vibrissae to solve the same task, see Subsection 2.1.3. But, the underlying actuation strategies are quite different. Micro vibrissae cannot rotate across a surface, only a translational displacement is possible because they lack appropriate efferent innervation and muscles. Assuming that rats detect a surface texture using their micro vibrissae, the benefit of analyzing a rotation of vibrissae in this context is doubtful. Consequently, in cooperation with biologists, the question which vibrissae is used for which task should be clarified first. Moreover, more details of the whisking strategy must be considered in the actuation of the artificial sensor. Since, on the neuronal level, there are multiple connections between sensor and motor nerves, the whisking strategy seems to be adaptive [Kleinfeld et al.; 2006]. Hence, a change in the scanning trajectory within a sweep will change the recorded signal significantly. In this context, the design of the artificial sensor used in the present work should be reconsidered including aspects like a closed-loop control regarding the sensor-actor information. A higher scanning velocity will be advantageous, too. This point is mainly restricted due to the vibrations induced by the motor of the guides. Here, the already mentioned structural refinements for an artificial follicle-sinus complex can be used to isolate the sensor from these vibrations. The choice of the reference frame for the support reactions is also influenced by the actuation strategy. This becomes important for the design of the artificial sensor and interpretation of the measured signals, e.g., as shown in Section A.3, a rotation of the sensor shaft can include different signal characteristics as a translational one.

All discussed points will improve the understanding of a natural vibrissa and provide design guidelines for an artificial sensor. Ideas for the application of a vibrissa-like sensor are discussed in Section 6.3.

6.3 Future application scenarios

In the following, three specific ideas for a possible, future application of a vibrissa-like tactile sensor are described.

Sensor for Grippers:
The impedance control of robots is still an object of research. Impedance control describes the interaction of a robot with its uncertain environment. Therefor, the forces as a result of contact or collision and the actual position are monitored continuously [Song et al.; 2017]. The interaction of a robot with its environment includes the manipulation of objects. Grasping of objects with robots is widely used in industry. There are many different concepts and underlying principles for grippers. Finger grippers can grasp an object by both a forced based connection or a tight fitted one. In either case, there are conditions regarding the grasping process. If an object gets clamped between two fingers of a gripper, the object and the fingers are related by a contact force that results in stress in the components and a friction force preventing slipping of the object out of the gripper. These effects are competing with each other. A larger load of the fingers on the object causes a larger friction force and consequently, a certain sticking of the object in the fingers. But at the same moment, a large load causes stress in the object which can be critical for the structural integrity. Therefore, an a priori knowledge about the object is necessary. If there is no a priori information about the object a sensory periphery is needed to ensure an adequate grasping of unknown objects. An array of vibrissa-like tactile sensors can be part of the sensory periphery. They can detect: where is the object and what is the object, as reported for the natural example, see Subsection 2.1.3. The robot/gripper can localize the object to avoid collisions as well as to do further scanning. The shape of the object can be recognized (see Section 5.1) to obtain the right configuration of the fingers in order to grasp the object by a tight fitted connection or the friction coefficient between vibrissa-like tactile sensor and object can be determined, compare Section 5.3. Furthermore, objects with similar shapes but different surface characteristics can be distinguished. All these tasks can be also done by other sensors but the integration of these tasks in one sensor is beneficial.

Pigging of pipes:
For example, in the petroleum industry, the maintenance of pipelines is a very common problem. The pipelines get cleaned and maintained by so-called pigs [Skerra; 2000].

Figure 6.1 – Example for an intelligent pig [Ogai, Bhattacharya; 2017].

A pig is a cylindrical object that goes through the pipeline, see Figure 6.1. The locomotion can be passive or active. If it is passive the pig is moved by the surrounding fluid else it is driven by several motors [Ogai, Bhattacharya; 2017]. There are intelligent pigs that observe the structural integrity of the pipeline including changes of the pipeline as deformations of the pipe and corrosion as well as wear of the inner lateral surface. These pigs are equipped with multiple sensors in order to fulfill the different scanning tasks.
The mentioned scanning tasks of a pig can be also fulfilled by an array of vibrissae-like tactile sensors. As demonstrated in the present work, a tactile sensor inspired by natural vibrissa can determine the shape of an object and detect information of its surface texture. Furthermore, both tasks can be performed simultaneously. The protection of the sensor by its own elasticity is valuable in this case, too, because it is possible that debris form unexpected large obstacles. Interestingly, there are already pigs equipped with bristles, but they are not used as sensors, just for cleaning. Consequently, updating these bristles with sensory elements in a way that they act like an array of vibrissae, various tasks regarding pigging can be combined in one sensor. Moreover, bristle-like structures can be also used for active locomotion in pipes as demonstrated in, e.g., [Becker et al.; 2013], so that the position of the pig is controllable.

ViBall:
The ViBall[6] is an idea/concept within the scope of soft robotics. Soft robotic designs are characterized by structural components that are flexible and with low stiffness [Campbell; 2018]. Following this concept, the idea of the ViBall is to combine sensor and actor in a single component. Proceeding from the objective of the present work, the sensor is inspired by natural vibrissae. For a combination of sensor and actor in one component, the effector and infector are the same structural component, compare Figure 2.7.

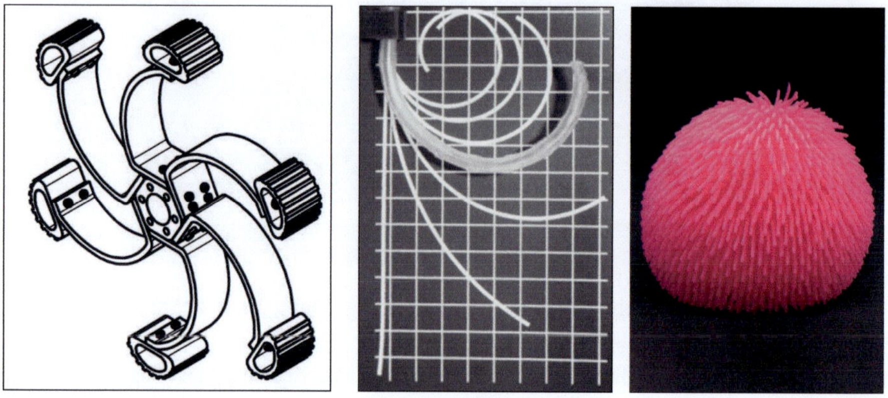

(a) Whegs **(b)** Fluid-mechanical actuator **(c)** Koosh ball

Figure 6.2 – In **(a)**- Whegs presented in [Stubbig et al.; 2017]; **(b)**- design of a fluid-mechnical actuator [Zentner et al.; 2019]; and **(c)**- Koosh ball toy.

Therefore, the effector must be a straight, slender, highly elastic structure. So-called Whegs are rimless wheels with a finite number of spokes [Stubbig et al.; 2017]. They are inspired by the locomotion of cockroaches [Quinn et al.; 2002] and allows for a terrestrial exploration of difficult terrain, see Figure 6.2(a). It seems to be possible to optimize the shape of a Whegs in a way that it fits in the present idea and, consequently, can serve as an effector. Whegs can be rigid or moderately stiff and are actuated by a motor, like classic wheels. Both facts are inconvenient for the present idea and must be changed. The authors of [Zentner et al.; 2019] designed an fluid-mechanical actor, see Figure 6.2(b). Here, the shape and stiffness of a slender, partly hollow, beam-like structure can be modified by controlling the internal pressure.

[6] ViBall is a syllabic abbreviation combing ViBrissa and Ball designed by the author

Combining a Whegs with the concept of a fluid-mechanical actor of [Zentner et al.; 2019], a tilting movement of the Whegs in order of shape and stiffness changes of the different spokes is conceivable.

The ViBall is the result of putting these three aspects together. Figure 6.2(c) illustrates the idea. As suggested, sensor and actor are combined in one component, here represented by slender, elastic, hair-like rods. Each sensor-actor unit can be controlled like the fluid-mechanical actuator of [Zentner et al.; 2019]. If there is no internal pressure and the base of these hair-like rods is equipped with force and torque sensors or others, all concepts mentioned in the present work can be used to explore an unknown object or terrain.

References

ADIBI, M.; DIAMOND, M. E.; ARABZADEH, E., 2012. Behavioral study of whisker-mediated vibration sensation in rats. *Proceedings of the National Academy of Sciences*. Vol. 109, pp. 971–976. Available from DOI: 10.1073/pnas.1116726109.

AHISSAR, E.; KNUTSEN, P., 2011. Vibrissal location coding. *Scholarpedia*. Vol. 6, pp. 6639. Available from DOI: 10.4249/scholarpedia.6639.

AHISSAR, E.; ORAM, T., 2015. Thalamic relay or cortico-thalamic processing? Old question, new answers. *Cerebral Cortex*. Vol. 25, pp. 845–848. Available from DOI: 10.1093/cercor/bht296.

AHL, A. S., 1986. The role of vibrissae in behavior: a status review. *Veterinary Research Communications*. Vol. 10, pp. 245–268. Available from DOI: 10.1007/BF02213989.

ALBARRACÍN, A. L.; FARFÁN, F. D.; FELICE, C. J.; DÉCIMA, E. E., 2006. Texture discrimination and multi-unit recording in the rat vibrissal nerve. *BMC Neuroscience*. Vol. 7, pp. 42. Available from DOI: 10.1186/1471-2202-7-42.

AMOLI, V.; KIM, J. S.; JEE, E.; CHUNG, Y. S.; KIM, S. Y.; KOO, J.; CHOI, H.; KIM, Y.; KIM, D. H., 2019. A bioinspired hydrogen bond-triggered ultrasensitive ionic mechanoreceptor skin. *Nature Communications*. Vol. 10, pp. 4019. Available from DOI: 10.1038/s41467-019-11973-5.

ARABZADEH, E.; VON HEIMENDAHL, M.; DIAMOND, M. E., 2009. Vibrissal texture decoding. *Scholarpedia*. Vol. 4, pp. 6640. Available from DOI: 10.4249/scholarpedia.6640.

ARABZADEH, E.; ZORZIN, E.; DIAMOND, M. E., 2005. Neuronal encoding of texture in the whisker sensory pathway. *PLoS Biology*. Vol. 3, pp. e17. Available from DOI: 10.1371/journal.pbio.0030017.

BAGDASARIAN, K.; SZWED, M.; KNUTSEN, P. M.; DEUTSCH, D.; DERDIKMAN, D.; PIETR, M.; SIMONY, E.; AHISSAR, E., 2013. Pre-neuronal morphological processing of object location by individual whiskers. *Nature Neuroscience*. Vol. 16, pp. 622–631. Available from DOI: 10.1038/nn.3378.

BALDEWEG, D.; WILL, C.; BEHN, C., 2014. Transversal vibrations of beams in context of vibrissae with foundations, discrete supports and various sections. In:

Proceedings of the 58th IWK, Ilmenau Scientific Colloquium: shaping the future by engineering. ilmedia, Available from urn:nbn:de:gbv:ilm1-2014iwk–081:7.

BECKER, F.; LYSENKO, V.; MINCHENYA, V.; ZEIDIS, I.; ZIMMERMANN, K., 2013. An approach to the dynamics of a vibration-driven robot. In: *Romansy 19 - Robot Design, Dynamics and Control.* Ed. by PFEIFFER, F.; RAMMERSTORFER, F. G.; SALENÇON, J.; SCHREFLER, B.; SERAFINI, P.; PADOIS, V.; BIDAUD, P.; KHATIB, O. Vienna: Springer, pp. 299–306. Available from DOI: 10.1007/978-3-7091-1379-0_37.

BEHN, C., 2013. *Mathematical modeling and control of biologically inspired uncertain motion systems with adaptive features.* Habilitation thesis. Technische Universität Ilmenau.

BEHN, C.; ACKERMANN, L.; WILL, C.; HELBIG, T.; STEIGENBERGER, J., 2017a. Vibrissa-based design of tapered tactile sensors for object sensing. In: *Proceedings of the 59th IWK, Ilmenau Scientific Colloquium: Engineering for a Changing World.* ilmedia, Available from urn:nbn:de:gbv:ilm1-2017iwk–050:5.

BEHN, C.; WILL, C.; SAUTER, A.; PREISS, T.; STEIGENBERGER, J., 2017b. Object sensing and shape detection using vibrissa hair-like sensors with intrinsic curvature. *International Journal on Advances in Intelligent Systems.* Vol. 10, pp. 179–188.

BELLI, H. M.; BRESEE, C. S.; GRAFF, M. M.; HARTMANN, M. J. Z., 2018. Quantifying the three-dimensional facial morphology of the laboratory rat with a focus on the vibrissae. *PLoS ONE.* Vol. 13, pp. e0194981. Available from DOI: 10.1371/journal.pone.0194981.

BELLI, H. M.; YANG, A. E. T.; BRESEE, C. S.; HARTMANN, M. J. Z., 2017. Variations in vibrissal geometry across the rat mystacial pad: base diameter, medulla, and taper. *Journal of Neurophysiology.* Vol. 4, pp. 1807–1820. Available from DOI: 10.1152/jn.00054.2016.

BERG, R. W.; KLEINFELD, D., 2003. Rhythmic whisking by rat: retraction as well as protraction of the vibrissae is under active muscular control. *Journal of Neurophysiology.* Vol. 89, pp. 104–117. Available from DOI: 10.1152/jn.00600.2002.

BERMEJO, R.; VYAS, A.; ZEIGLER, H. P., 2002. Topography of rodent whisking - I. Two-dimensional monitoring of whisker movements. *Somatosensory & Motor Research.* Vol. 19, pp. 341–346. Available from DOI: 10.1080/0899022021000037809.

BHUSHAN, B.; WEI, G.; HADDAD, P., 2005. Friction and wear studies of human hair and skin. *Wear*. Vol. 259, pp. 1012–1021. Available from DOI: 10.1016/j.wear.2004.12.026.

BIRDWELL, J. A.; SOLOMON, J. H.; THAJCHAYAPONG, M.; TAYLOR, M. A.; CHEELY, M.; TOWAL, R. B.; CONRADT, J.; HARTMANN, M. J. Z., 2007. Biomechanical models for radial distance determination by the rat vibrissal system. *Journal of Neurophysiology*. Vol. 98, pp. 2439–2455. Available from DOI: 10.1152/jn.00707.2006.

BMWI, 2018. *Industrie 4.0* [online] [visited on 2019-12-10]. Available from: https://www.bmwi.de/Redaktion/EN/Dossier/industrie-40.html.

BOSMAN, L. W. J. et al., 2011. Anatomical pathways involved in generating and sensing rhytmic whisker movements. *Frontiers in Behavioral Neuroscience*. Vol. 5, pp. 53. Available from DOI: 10.3389/fnint.2011.00053.

BOUBENEC, Y.; CLAVERIE, L. N.; SHULZ, D. E.; DEBRÉGEAS, G., 2014. An amplitude modulation/demodulation scheme for whisker-based texture perception. *The Journal of Neuroscience*. Vol. 34, pp. 10832–10843. Available from DOI: 10.1523/JNEUROSCI.0534-14.2014.

BOUBENEC, Y.; SHULZ, D. E.; DEBRǴEAS, G., 2012. Whisker encoding of mechanical events during active tactile exploration. *Frontiers in Behavioral Neuroscience*. Vol. 6, pp. article 74. Available from DOI: 10.3389/fnbeh.2012.00074.

BRECHT, M.; PREILOWSKI, B.; MERZENICH, M. M., 1997. Functional architecture of the mystacial vibrissae. *Behavioural Brain Research*. Vol. 84, pp. 81–97. Available from DOI: 10.1016/S0166-4328(97)83328-1.

BUSH, N. E.; SOLLA, S. A.; HARTMANN, M. J. Z., 2016. Whisking mechanics and active sensing. *Current Opinion in Neurobiology*. Vol. 40, pp. 178–188. Available from DOI: 10.1016/j.conb.2016.08.001.

CAMPAGNER, D.; EVANS, M. H.; LOFT, M. S. E.; PETERSEN, R. S., 2018. What the whiskers tell the brain. *Neuroscience*. Vol. 368, pp. 95–108. Available from DOI: 10.1016/j.neuroscience.2017.08.005.

CAMPBELL, S., 2018. The robotics revolution will be soft: soft robotics proliferate-along with their sources of inspiration. *IEEE Pulse*. Vol. 9, pp. 19–24. Available from DOI: 10.1109/MPUL.2018.2814240.

CARL, K., 2009. *Technische Biologie des Tasthaar-Sinnessystems als Gestaltungsgrundlage für taktile stiftführende Mechanosensoren.* PhD thesis. Technische Universität Ilmenau.

CARL, K.; HILD, W.; MAMPEL, J.; SCHILLING, C.; UHLIG, R.; WITTE, H., 2012. Characterization of statical properties of rat's whisker system. *IEEE Sensors Journal.* Vol. 12, pp. 340–349. Available from DOI: 10.1109/JSEN.2011.2114341.

CARL, K.; SCHILLING, C.; WITTE, H., 2008. Der Reizleitungsapparat: Eine technisch motivierte Theorie der Sinnesorgane. *Mechatronik.* Vol. 116, pp. 102–109.

CARVELL, G. E.; SIMONS, D. J., 1990. Biometric analyses of vibrissal tactile discrimination in the rat. *The Journal of Neuroscience.* Vol. 10, pp. 2638–2648. Available from DOI: 10.1523/JNEUROSCI.10-08-02638.1990.

CARVELL, G. E.; SIMONS, D. J., 1995. Task- and subject- related differences in sensorimotor behavior during active touch. *Somatosensory & Motor Research.* Vol. 12, pp. 1–9. Available from DOI: 10.3109/08990229509063138.

CARVELL, G. E.; SIMONS, D. J., 2017. Effect of whisker geometry on contact force produced by vibrissae moving at different velocities. *Journal of Neurophysiology.* Vol. 118, pp. 1637–1649. Available from DOI: 10.1152/jn.00046.2017.

CHEUNG, J.; MAIRE, P.; KIM, J.; SY, J.; HIRES, S. A., 2019. The sensorimotor basis of whisker-guided anteroposterior object localization in head-fixed mice. *Current Biology.* Vol. 29, pp. 3029–3040. Available from DOI: 10.1016/j.cub.2019.07.068.

CLAVERIE, L. N.; BOUBENEC, Y.; DEBRǴEAS, G.; PREVOST, A. M.; WANDERSMAN, E., 2017. Whisker contact detection of rodents based on slow and fast mechanical inputs. *Frontiers in Behavioral Neuroscience.* Vol. 10, pp. article 251. Available from DOI: 10.3389/fnbeh.2016.00251.

CLEMENTS, T. N.; RAHN, C. D., 2005. Three-dimensional contact imaging with an actuated whisker. In: *2005 IEEE/RSJ International Workshop on Intelligent Robots and Systems.* IEEE, pp. 598–603. Available from DOI: 10.1109/IROS.2005.1545213.

CURTIS, J. C.; KLEINFELD, D., 2006. Seeing what the mouse sees with its vibrissae: a matter of behavioral state. *Neuron.* Vol. 50, pp. 524–526. Available from DOI: 10.1016/j.neuron.2006.05.004.

CUTKOSKY, M. R.; HOWE, R. D.; PROVANCHER, W. R., 2008. Force and tactile sensors. In: *Springer Handbook of Robotics.* Ed. by SICILIANO, B.; KHATIB, O.

Berlin, Heidelberg: Springer, pp. 455–476. Available from DOI: 10.1007/978-3-540-30301-5_20.

DAVIM, J. P., 2010. *Surface integrity in machining.* London: Springer. ISBN 978-1-84882-873-5.

DEHNHARDT, G.; MAUCK, B.; HYVÄRINEN, H., 1998. Ambient temperature does not affect the tactile sensitivity of mystacial vibrissae in harbour seals. *The Journal of Experimental Biology.* Vol. 201, pp. 3023–3029.

DEUTSCH, S.; DEUTSCH, A., 1993. *Understanding the nervous system: An engineering perspective.* Piscataway, NJ: Wiley-IEEE Press. ISBN 978-0879422967.

DIAMOND, M. E., 2010. Texture sensation through the fingertips and the whiskers. *Current Opinion in Neurobiology.* Vol. 20, pp. 319–327. Available from DOI: 10.1016/j.conb.2010.03.004.

DIAMOND, M. E.; ARABZADEH, E., 2013. Whisker sensory system - from receptor to decision. *Progress in Neurobiology.* Vol. 103, pp. 28–40. Available from DOI: 10.1016/j.pneurobio.2012.05.013.

DIAMOND, M. E.; VON HEIMENDAHL, M.; ARABZADEH, E., 2008a. Whisker-mediated texture discrimination. *PLoS Biology.* Vol. 6, pp. e220. Available from DOI: 10.1371/journal.pbio.0060220.

DIAMOND, M. E.; VON HEIMENDAHL, M.; KNUTSEN, P. M.; KLEINFELD, D.; AHISSAR, E., 2008b. 'Where' and 'what' in the whisker sensorimotor system. *Nature reviews. Neuroscience.* Vol. 9, pp. 601–612. Available from DOI: 10.1038/nrn2411.

DÖRFL, J., 1982. The musculature of the mystacial vibrissae of the white mouse. *Journal of Anatomy.* Vol. 135, pp. 147–154.

DRESIG, H.; HOLZWEISSIG, F., 2009. *Maschinendynamik.* Berlin: Springer.

EBARA, S.; FURUTA, T.; KUMAMOTO, K., 2017. Vibrissal mechanoreceptors. *Scholarpedia.* Vol. 12, pp. 32372. Available from DOI: 10.4249/scholarpedia.32372.

EBARA, S.; KUMAMOTO, K.; MATSUURA, T.; MAZURKIEWICZ, J. E.; RICE, F. L., 2002. Similarities and differences in the innervation of mystacial vibrissal follicle-sinus complexes in the rat and cat: a confocal microscopic study. *The Journal of Comparative Neurology.* Vol. 449, pp. 103–119. Available from DOI: 10.1002/cne.10277.

EGO-STENGEL, V.; ABBASI, A.; LARROCHE, M.; LASSAGNE, H.; BOUBENEC, Y.; SHULZ, D. E., 2019. Mechanical coupling through the skin affects whisker movements and tactile information encoding. *Journal of Neurophysiology*. Vol. 122, pp. 1606–1622. Available from DOI: 10.1152/jn.00863.2018.

ENGER, M.; FEINLE, P., 2014. Stick-Slip, Testing Methods. In: *Encyclopedia of Lubricants and Lubrication*. Ed. by MANG, T. Berlin, Heidelberg: Springer, pp. 1988–1997. Available from DOI: 10.1007/978-3-642-22647-2_490.

ERDSACK, N.; DEHNHARDT, G.; HANKE, W., 2014. Thermoregulation of the vibrissal system in harbor seals (Phoca vitulina) and cape fur seals (Arctocephalus pusillus pusillus). *Journal of Experimental Marine Biology and Ecology*. Vol. 452, pp. 111–118. Available from DOI: 10.1016/j.jembe.2013.12.011.

ESTEBANEZ, L.; FÉRŹOU, I.; EGO-STENGEL, V.; SHULZ, D. E., 2018. Representation of tactile scenes in the rodent barrel cortex. *Neuroscience*. Vol. 368, pp. 81–94. Available from DOI: 10.1016/j.neuroscience.2017.08.039.

EVANS, M. H.; LOFT, M. S. E.; CAMPAGNER, D.; PETERSEN, R. S., 2019. Sensing the environment with whiskers. *Oxford Research Encyclopedia of Neuroscience*, pp. 1–39. Available from DOI: 10.1093/acrefore/9780190264086.013.226.

FARFÁN, F. D.; ALBARRACÍN, A. L.; FELICE, C. J., 2011. Electrophysiological characterization of texture information slip-resistance dependent in the rat vibrissal nerve. *BMC Neuroscience*. Vol. 12, pp. 32. Available from DOI: 10.1186/1471-2202-12-32.

FASSIHI, A.; AKRAMI, A.; PULECCHI, F.; SCHÖNFELDER, V.; DIAMOND, M. E., 2017. Transformation of perception from sensory to motor cortex. *Current Biology*. Vol. 27, pp. 1–12. Available from DOI: 10.1016/j.cub.2017.05.011.

FASSIHI, A.; ESMAEILI, V.; AKRAMI, A.; MANZINO, F.; DIAMOND, M. E., 2015. Sensation of a „noisy" whisker vibration in rats. In: *Biomimetic and biohybrid systems*. Ed. by PRESCOTT, T. J.; LEPORA, N. F.; MURA, A.; VERSCHURE, P. F. M. J. Berlin, Heidelberg: Springer, pp. 341–342. Available from DOI: 10.1007/978-3-642-31525-1_33.

FEND, M.; BOVET, S.; YOKOI, H.; PFEIFER, R., 2003. An active artificial whisker array for texture discrimination. In: *2003 IEEE/RSJ International Workshop on Intelligent Robots and Systems*. IEEE, pp. 1044–1049. Available from DOI: 10.1109/IROS.2003.1248782.

FURUTA, T.; BUSH, N. E.; YANG, A.; EBARA, S.; MIYAZAKI, N.; MURATA, K.; HIRAI, D.; SHIBATA, K.; HARTMANN, M. J. Z., 2020. The cellular and mechanical basis for response characteristics of identified primary afferents in the rat vibrissal system. *Current Biology*. Vol. in press. Available from DOI: 10.1016/j.cub.2019.12.068.

GANGULY, K.; KLEINFELD, D., 2004. Goal-directed whisking increases phase-locking between vibrissa movement and electrical activity in primary sensory cortex in rat. *Proceedings of the National Academy of Sciences of the United States of America*. Vol. 101, pp. 12348–12353. Available from DOI: 10.1073/pnas.0308470101.

GARRETT, B. S., 2009. It's not you, it's me. Really. A subset of neurons in rat barrel cortex integrate information about the object a whisker contacts with the motion of the whisker at the time of contact, setting the stage for a highly specialized object localization system. *Nature Neuroscience*. Vol. 12, pp. 374–375. Available from DOI: 10.1038/nn0409-374.

GRANT, R. A.; ITSKOV, P. M.; TOWAL, R. B.; PRESCOTT, T. J., 2014. Active touch sensing: finger tips, whiskers, and antennae. *Frontiers in Behavioral Neuroscience*. Vol. 8, pp. 50. Available from DOI: 10.15406/ijbsbe.2017.02.00030.

GRANT, R. A.; MITCHINSON, B.; FOX, C. W.; PRESCOTT, T. J., 2008. Active touch sensing in the rat: anticipatory and regulatory control of whisker movements during surface exploration. *Journal of Neurophysiology*. Vol. 101, pp. 862–874. Available from DOI: 10.1152/jn.90783.2008.

GRANT, R. A.; SPERBER, A. L.; PRESCOTT, T. J., 2012. The role of orienting in vibrissal touch sensing. *Frontiers in Behavioral Neuroscience*. Vol. 6, pp. 39. Available from DOI: 10.3389/fnbeh.2012.00039.

GUIĆ-ROBLES, E.; VALDIVIESO, C.; GUAJARDO, G., 1989. Rats can learn a roughness discrimination using only their vibrissal system. *Behavioural Brain Research*. Vol. 31, pp. 285–289. Available from DOI: 10.1016/0166-4328(89)90011-9.

HAIDARLIU, S., 2015. Whisking musculature. *Scholarpedia*. Vol. 10, pp. 32331. Available from DOI: 10.4249/scholarpedia.32331.

HAIDARLIU, S.; BAGDASARIAN, K.; SHINDE, N.; AHISSAR, E., 2017. Muscular basis of whisker torsion in mice and rats. *The Anatomical Record: Advances in Integrative Anatomy and Evolutionary Biology*. Vol. 300, pp. 1643–1653. Available from DOI: 10.1002/ar.23623.

HARTMANN, M. J. Z., 2001. Active sensing capabilities of the rat whisker system. *Autonomous Robots*. Vol. 11, pp. 249–254. Available from DOI: 10.1023/A:1012439023425.

HARTMANN, M. J. Z., 2009. Active touch, exploratory movements, and sensory prediction. *Integrative and Comparative Biology*. Vol. 49, pp. 681–690. Available from DOI: 10.1093/icb/icp107.

HARTMANN, M. J. Z., 2011. A night in the life of a rat: vibrissal mechanics and tactile exploration. *Annals of the New York Academy of Sciences*. Vol. 1225, pp. 110–118. Available from DOI: 10.1111/j.1749-6632.2011.06007.x.

HARTMANN, M. J. Z., 2015. Vibrissa mechanical properties. *Scholarpedia*. Vol. 10, pp. 6636. Available from DOI: 10.4249/scholarpedia.6636.

HARTMANN, M. J. Z.; JOHNSON, N. J.; TOWAL, R. B.; ASSAD, C., 2003. Mechanical characteristics of rat vibrissae: resonant frequencies and damping in isolated whiskers and in the awake behaving animal. *The Journal of Neuroscience*. Vol. 23, pp. 6510–6519. Available from DOI: 10.1523/JNEUROSCI.23-16-06510.2003.

HARVEY, M. A.; BERMEJO, R.; ZEIGLER, H. P., 2001. Discriminative whisking in the head-fixed rat: optoelectronic monitoring during tactile detection and discrimination tasks. *Somatosensory & Motor Research*. Vol. 18, pp. 211–222. Available from DOI: 10.1080/01421590120072204.

HIPP, J.; ARABZADEH, E.; ZORZIN, E.; CONRADT, J.; KAYSER, C.; DIAMOND, M. E.; KÖNIG, P., 2006. Texture signals in whisker vibrations. *Journal of Neurophysiology*. Vol. 95, pp. 1792–1799. Available from DOI: 10.1152/jn.01104.2005.

HIRES, S. A.; PAMMER, L.; SVOBODA, K.; GOLOMB, D., 2013. Tapered whiskers are required for active tactile sensation. *eLife*. Vol. 2, pp. e01350. Available from DOI: 10.7554/eLife.01350.

HIRES, S. A.; SCHUYLER, A.; SY, J.; HUANG, V.; WYCHE, I.; WANG, X.; GOLOMB, D., 2016. Beyond cones: an improved model of whisker bending based on measured mechanics and tapering. *Journal of Neurophysiology*. Vol. 116, pp. 812–824. Available from DOI: 10.1152/jn.00511.2015.

HIROSE, S.; INOUE, S.; YONEDA, K., 1989. The whisker sensor and the transmission of multiple sensor signals. *Advanced Robotics*. Vol. 4, pp. 105–117. Available from DOI: 10.1163/156855390X00099.

HOBBS, J. A.; TOWAL, R. B.; HARTMANN, M. J. Z., 2016. Evidence for functional groupings of vibrissae across the rodent mystacial pad. *PLoS Computational Biology*. Vol. 12, pp. e1004109. Available from DOI: 10.1371/journal.pcbi.1004109.

HOREV, G.; SAIG, A.; KNUTSEN, P. M.; PIETR, M.; YU, C.; AHISSAR, E., 2011. Motor-sensory convergence in object localization: a comparative study in rats and humans. *Current Biology*. Vol. 366, pp. 3070–3076. Available from DOI: 10.1098/rstb.2011.0157.

HU, X.; URHIE, O.; CHANG, K.; HOSTETLER, R.; AGMON, A., 2018. A novel method for training mice in visuo-tactile 3-D object discrimination and recognition. *Frontiers in Behavioral Neuroscience*. Vol. 12, pp. article 274. Available from DOI: 10.3389/fnbeh.2018.00274.

HUET, L. A.; RUDNICKI, J. W.; HARTMANN, M. J. Z., 2017. Tactile sensing with whiskers of various shapes: determining the three-dimensional location of object contact based on mechanical signals at the whisker base. *Soft Robotics*. Vol. 4, pp. 88–102. Available from DOI: 10.1089/soro.2016.0028.

IOUGUINA, A., 2012. *Bionics ≠ Biomimetics ≠ Biomimicry* [online] [visited on 2019-12-10]. Available from: https://biologytodesign.wordpress.com/2012/05/08/design-biology-linguistics/.

ISETT, B. R.; FEASEL, S. H.; LANE, M. A.; FELDMAN, D. E., 2018. Slip-based coding of local shape and texture in mouse S1. *Neuron*. Vol. 97, pp. 418–433.e5. Available from DOI: 10.1016/j.neuron.2017.12.021.

JADHAV, S. P.; FELDMAN, D. E., 2010. Texture coding in the whisker system. *Current Opinion in Neurobiology*. Vol. 20, pp. 313–318. Available from DOI: 10.1016/j.conb.2010.02.014.

JENKS, R. A.; VAZIRI, A.; BOLOORI, A.; STANLEY, G. B., 2010. Self-motion and the shaping of sensory signals. *Journal of Neurophysiology*. Vol. 103, pp. 2195–2207. Available from DOI: 10.1152/jn.00106.2009.

JOHANASTROM, K.; CANUDAS-DE-WIT, C., 2008. Revisiting the LuGre friction model. *IEEE Control Systems Magazine*. Vol. 28, pp. 101–114. Available from DOI: 10.1109/MCS.2008.929425.

KAN, Q.; RAJAN, R.; FU, J.; KANG, G.; YAN, W., 2013. Elastic modulus of rat whiskers - a key biomaterial in the rat whisker sensory system. *Materials Research Bulletin*. Vol. 48, pp. 5026–5032. Available from DOI: 10.1016/j.materresbull.2013.04.070.

KEREKES, P.; DARET, A.; SHULZ, D. E.; EGO-STENGEL, V., 2017. Bilateral discrimination of tactile patterns without whisking in freely running rats. *The Journal of Neuroscience*. Vol. 37, pp. 7567–7579. Available from DOI: 10.1523/JNEUROSCI.0528-17.2017.

KIM, D.; MÖLLER, R., 2007. Biomimetic whiskers for shape recognition. *Robotics and Autonomous Systems*. Vol. 55, pp. 229–243. Available from DOI: 10.1016/j.robot.2006.08.001.

KIM, J.; YOO, J.; LEE, J.; KOH, K.; SONG, W., 2012. A mechanism of rat vibrissal movement based on actual morphology of the intrinsic muscle using three-dimensional reconstruction. *Cells Tissues Organs*. Vol. 196, pp. 565–569. Available from DOI: 10.1159/000338332.

KLAUER, G. J., 2001. *Vibrissen: Analyse eines taktilen Sinnesorgans Band 1*. Habilitation thesis. Universität Gesamthochschule Essen.

KLEINFELD, D.; AHISSAR, E.; DIAMOND, M. E., 2006. Active sensation: insights from the rodent vibrissa sensorimotor system. *Current Opinion in Neurobiology*. Vol. 16, pp. 435–444. Available from DOI: 10.1016/j.conb.2006.06.009.

KLEINFELD, D.; DESCHÊNES, M.; MOORE, J. D., 2015. The central pattern generator for rhythmic whisking. In: *Sensorimotor integration in the whisker system*. Ed. by KRIEGER, P.; GROH, A. New York, NY: Springer, pp. 149–165. Available from DOI: 10.1007/978-1-4939-2975-7_7.

KNUTSEN, P. M.; AHISSAR, E., 2009. Orthogonal coding of object location. *Trends in Neurosciences*. Vol. 32, pp. 101–109. Available from DOI: 10.1016/j.tins.2008.10.002.

KNUTSEN, P. M.; BIESS, A.; AHISSAR, E., 2008. Vibrissal kinematics in 3D: tight coupling of azimuth, elevation, and torsion across different whisking modes. *Neuron*. Vol. 59, pp. 35–42. Available from DOI: 10.1016/j.neuron.2008.05.013.

KRIEGER, P.; GROH, A. (eds.), 2015. *Sensorimotor integration in the whisker system*. New York, NY: Springer.

KRYSTEK, M., 2009. ISO filters for precision engineering. *Technisches Messen*. Vol. 76, pp. 133–159. Available from DOI: 10.1524/teme.2009.0927.

KURUPPATH, P.; GUGIG, E.; AZOUZ, R., 2014. Microvibrissae-based texture discrimination. *The Journal of Neuroscience*. Vol. 34, pp. 5115–5120. Available from DOI: 10.1523/JNEUROSCI.4217-13.2014.

LEDERMAN, S. J.; KLATZKY, R. L., 1987. Hand movements: a window into haptic object recognition. *Cognitive Psychology*. Vol. 19, pp. 342–368. Available from DOI: 10.1016/0010-0285(87)90008-9.

LEISER, S. C.; MOXON, K. A., 2007. Responses of trigeminal ganglion neurons during natural whisking behaviors in the awake rat. *Neuron*. Vol. 53, pp. 117–133. Available from DOI: 10.1016/j.neuron.2006.10.036.

LEPORA, N. F.; VERSCHURE, P.; PRESCOTT, T. J., 2013. The state of the art in biomimetics. *Bioinspiration & Biomimetics*. Vol. 8, pp. 013001. Available from DOI: 10.1088/1748-3182/8/1/013001.

LIN, D. C.; MCGOWAN, C. P.; BLUM, K. P.; TING, L. H., 2019. Yank: the time derivative of force is an important biomechanical variable in sensorimotor systems. *Journal of Experimental Biology*. Vol. 222, pp. Pt 18. Available from DOI: 10.1242/jeb.180414.

MARANGONI, R. R.; SCHLEICHERT, J.; RAHNEBERG, I.; MASTYLO, R.; MANSKE, E.; FRÖHLICH, T., 2017. Multi-component force measurement in micromachining. *Technisches Messen*. Vol. 84, pp. 587–592. Available from DOI: 10.1515/teme-2016-0066.

MARAVALL, M.; DIAMOND, M. E., 2015. Functional principles of whisker-mediated touch perception. In: *Sensorimotor integration in the whisker system*. Ed. by KRIEGER, P.; GROH, A. New York, NY: Springer, pp. 169–193. Available from DOI: 10.1007/978-1-4939-2975-7_8.

MERKER, L.; FISCHER CALDERON, S. J.; SCHARFF, M.; ALENCASTRE MIRANDA, J. H.; BEHN, C., 2020. Effects of multi-point contacts during object contour scanning using a biologically-inspired tactile sensor. *Sensors*. Vol. 20, pp. 2077. Available from DOI: 10.3390/s20072077.

MERKER, L.; STEIGENBERGER, J.; BEHN, C., 2019. 3D recognition of obstacles using a vibrissa-like tactile sensor. In: *2019 IEEE International Conference on Flexible and Printable Sensors and Systems*. IEEE, pp. 1–3. Available from DOI: 10.1109/FLEPS.2019.8792255.

MERKER, L.; WILL, C.; STEIGENBERGER, J.; BEHN, C., 2018. Object shape recognition and reconstruction using pivoted tactile sensors. *Mathematical Problems in Engineering*. Vol. 2018, pp. Article ID 1613945. Available from DOI: 10.1155/2018/1613945.

MITCHINSON, B.; GURNEY, K. N.; REDGRAVE, P.; MELHUISH, C.; PIPE, A. G.; PEARSON, M.; GILHESPY, I.; PRESCOTT, T. J., 2004. Empirically inspired simulated electro-mechanical model of the rat mystacial follicle-sinus complex. *Proceedings of the Royal Society B: Biological Sciences*. Vol. 271, pp. 2509–2516. Available from DOI: 10.1098/rspb.2004.2882.

MITCHINSON, B.; MARTIN, C. J.; GRANT, R. A.; PRESCOTT, T. J., 2007. Feedback control in active sensing: rat exploratory whisking is modulated by environmental contact. *Proceedings of the Royal Society B: Biological Sciences*. Vol. 274, pp. 1035–1041. Available from DOI: 10.1098/rspb.2006.0347.

MITCHINSON, B.; PEARSON, M. J.; PIPE, A. G.; PRESCOTT, T. J., 2014. Biomimetic tactile target acquisition, tracking and capture. *Robotics and Autonomous Systems*. Vol. 62, pp. 366–375. Available from DOI: 10.1016/j.robot.2013.08.013.

MOORE, C. I.; ANDERMANN, M. L., 2005. The vibrissa resonance hypothesis. In: *Neural plasticity in adult somatic sensory-motor systems*. Ed. by EBNER, F. F. Boca Raton, FL: Taylor & Francis, pp. 21–60.

NACHTIGALL, W., 2002. *Bionik: Grundlagen und Beispiele für Ingenieure und Naturwissenschaftler*. Berlin, Heidelberg: Springer. ISBN 978-3-642-62399-8.

NEIMARK, M. A.; ANDERMANN, M. L.; HOPFIELD, J. J.; MOORE, C. I., 2003. Vibrissa resonance as a transduction mechanism for tactile encoding. *The Journal of Neuroscience*. Vol. 23, pp. 6499–6509. Available from DOI: 10.1523/JNEUROSCI.23-16-06499.2003.

OGAI, H.; BHATTACHARYA, B., 2017. *Pipe inspection robots for structural health and condition monitoring*. New Dehli: Spinger.

OLADAZIMI, M.; BRENDEL, W.; SCHWARZ, C., 2018. Biomechanical texture coding in rat whiskers. *Scientific Reports*. Vol. 8, pp. 11139. Available from DOI: 10.1038/s41598-018-29225-9.

PACEJKA, H. B.; BAKKER, E., 1992. The magic formula tyre model. *Vehicle System Dynamics*. Vol. 21, pp. 1–18. Available from DOI: 10.1080/00423119208969994.

PAMMER, L.; O'CONNOR, D. H.; HIRES, S. A.; CLACK, N. G.; HUBER, D.; MYERS, E. W.; SVOBODA, K., 2013. The mechanical variables underlying object localization along the axis of the whisker. *The Journal of Neuroscience*. Vol. 33, pp. 6726–6741. Available from DOI: 10.1523/JNEUROSCI.4316-12.2013.

PEARSON, M. J.; MITCHINSON, B.; SULLIVAN, J. C.; PIPE, A. G.; PRESCOTT, T. J., 2011. Biomimetic vibrissal sensing for robots. *Philosophical Transactions of the Royal Society B: Biological Sciences*. Vol. 366, pp. 3085–3096. Available from DOI: 10.1098/rstb.2011.0164.

POPOV, V. L., 2010. *Contact mechanics and friction: physical principles and applications*. Berlin: Springer. ISBN 978-3-642-10802-0.

POPP, K.; PANNING, L.; SEXTRO, W., 2016. Vibration damping by friction forces: theory and applications. *Modal Analysis*. Vol. 9, pp. 419–448. Available from DOI: 10.1177/107754603030780.

PRESCOTT, T. J.; MITCHINSON, B.; GRANT, R. A., 2011. Vibrissal behavior and function. *Scholarpedia*. Vol. 6, pp. 6642. Available from DOI: 10.4249/scholarpedia.6642.

PRESCOTT, T. J.; PEARSON, M. J.; MITCHINSON, B.; SULLIVAN, J. C. W.; PIPE, A., 2009. Whisking with robots: from rat vibrissae to biomimetic technology for active touch. *IEEE Robotics & Automation Magazine*. Vol. 16, pp. 42–50. Available from DOI: 10.1109/MRA.2009.933624.

QUINN, R. D.; OFFI, J. T.; KINGSLEY, D. A.; RITZMANN, R. E., 2002. Improved mobility through abstracted biological principles. In: *2002 IEEE/RSJ International Workshop on Intelligent Robots and Systems*. IEEE, pp. 2652–2657. Available from DOI: 10.1109/IRDS.2002.1041670.

QUIST, B. W.; FARUQI, R. A.; HARTMANN, M. J. Z., 2011. Variation in Young's modulus along the length of a rat vibrissa. *Journal of Biomechanics*. Vol. 44, pp. 2775–2781. Available from DOI: 10.1016/j.jbiomech.2011.08.027.

QUIST, B. W.; HARTMANN, M. J. Z., 2012. Mechanical signals at the base of a rat vibrissa: the effect of intrinsic vibrissa curvature and implications for tactile exploration. *Journal of Neurophysiology*. Vol. 107, pp. 2298–2312. Available from DOI: 10.1152/jn.00372.2011.

QUIST, B. W.; SEGHETE, V.; HUET, L. A.; MURPHEY, T. D.; HARTMANN, M. J. Z., 2014. Modeling forces and moments at the base of a rat vibrissa during noncontact whisking and whisking against an object. *The Journal of Neuroscience*. Vol. 34, pp. 9828–9844. Available from DOI: 10.1523/JNEUROSCI.1707-12.2014.

RANGELOW, I. W.; IVANOV, T.; AHMAD, A.; KAESTNER, M.; LENK, C.; BOZCHALOOI, I. S.; XIA, F.; YOUCEF-TOUMI, K.; HOLZ, M.; REUM, A., 2017. Active scanning probes: a versatile toolkit for fast imaging and emerging nanofabrication. *Journal of Vacuum Science & Technology B*. Vol. 35, pp. 06G101. Available from DOI: 10.1116/1.4992073.

RICE, F. L., 1993. Structure, vascularization, and innervation of the mystacial pad of the rat as revealed by the lectin Griffonia simplicifolia. *The Journal of Comparative Neurology*. Vol. 337, pp. 386–399. Available from DOI: 10.1002/cne.903370304.

RICE, F. L.; MANCE, A.; MUNGER, B. L., 1986. A comparative light microscopic analysis of the sensory innervation of the mystacial pad. I. Innervation of vibrissal follicle-sinus complexes. *The Journal of Comparative Neurology*. Vol. 252, pp. 154–174. Available from DOI: 10.1002/cne.902520203.

RITT, J. T.; ANDERMANN, M. L.; MOORE, C. I., 2008. Embodied information processing: vibrissa mechanics and texture features shape micromotions in actively sensing rats. *Neuron*. Vol. 57, pp. 599–613. Available from DOI: 10.1016/j.neuron.2007.12.024.

ROBBINS, C. R., 2012. *Chemical and physical behavior of human Hair*. Berlin, Heidelberg: Springer. ISBN 978-3-642-25610-3.

ROONEY, T.; PEARSON, M. J.; PIPE, A. G., 2015. Measuring the local viscosity and velocity of fluids using a biomimetic tactile whisker. In: *Biomimetic and Biohybrid Systems*. Ed. by WILSON, S. P.; VERSCHURE, P. F. M. J.; MURA, A.; PRESCOTT, T. J. Cham: Springer International Publishing, pp. 75–85. Available from DOI: 10.1007/978-3-319-22979-9_7.

RUBINSTEIN, S. M.; COHEN, G.; FINEBERG, J., 2009. Visualizing stick–slip: experimental observations of processes governing the nucleation of frictional sliding. *Journal of Physics D: Applied Physics*. Vol. 42, pp. 214016. Available from DOI: 10.1088/0022-3727/42/21/214016.

SARAF-SINIK, I.; ASSA, E.; AHISSAR, E., 2015. Motion makes sense: an adaptive motor-sensory strategy underlies the perception of object location in rats. *The Journal of Neuroscience*. Vol. 35, pp. 8777–8789. Available from DOI: 10.1523/JNEUROSCI.4149-14.2015.

SCHARFF, M.; ALENCASTRE, J. H.; BEHN, C., 2019a. Detection of surface texture with an artificial tactile sensor. In: *Interdisciplinary Applications of Kinematics*. Ed. by KECSKEMÉTHY, A.; GEU FLORES, F.; CARRERA, E.; ELIAS, D. A. Cham: Springer, pp. 43–50. Available from DOI: 10.1007/978-3-030-16423-2_4.

SCHARFF, M.; ALENCASTRE, J. H.; WITTE, H.; ZIMMERMANN, K.; STEIGENBERGER, J.; BEHN, C., 2017a. Investigations on the mechanical relevance of prominent vibrissa features for surface texture detection. In: *Proceedings of the 59th IWK, Ilmenau Scientific Colloquium: Engineering for a Changing World*. ilmedia, Available from urn:nbn:de:gbv:ilm1-2017iwk–094:0.

SCHARFF, M.; BEHN, C.; STEIGENBERGER, J.; ALENCASTRE, J. H., 2016. Towards the development of tactile sensors for surface texture detection. In: *Proceedings of the Fifth International Conference on Intelligent Systems and Applications*. IARIA, pp. 33–38. Available from Article: 2_20_60040.

SCHARFF, M.; DARNIEDER, M.; STEIGENBERGER, J.; ALENCASTRE, J. H.; BEHN, C., 2018. Theoretical investigations on the behavior of artificial sensors for surface texture detection. In: *Dynamical Systems in Theoretical Perspective. DSTA 2017. Springer Proceedings in Mathematics & Statistics.* Ed. by AWREJCEWICZ, J. Cham: Springer, pp. 311–321. Available from DOI: 10.1007/978-3-319-96598-7_25.

SCHARFF, M.; DARNIEDER, M.; STEIGENBERGER, J.; BEHN, C., 2017b. Towards the development of tactile sensors for determination of static friction coefficient to surfaces. In: *Microactuators and Micromechanisms. Mechanisms and Machine Science.* Ed. by ZENTNER, L.; CORVES, B.; JENSEN, B.; LOVASZ, E. C. Cham: Springer, pp. 39–48. Available from DOI: 10.1007/978-3-319-45387-3_4.

SCHARFF, M.; SCHORR, P.; BECKER, T.; RESAGK, C.; ALENCASTRE MIRANDA, J. H.; BEHN, C., 2019b. An artificial vibrissa-like sensor for detection of flows. *Sensors.* Vol. 19, pp. 3892. Available from DOI: 10.3390/s19183892.

SCHARFF, M.; STEIGENBERGER, J.; BEHN, C.; ALENCASTRE, J., 2017c. Influence of vibrissa morphology on artificial tactile sensors for surface texture detection. *Proccedings in Applied Mathematics and Mechanics.* Vol. 17, pp. 219–220. Available from DOI: 10.1002/pamm.201710079.

SCHMIDT, M.; WITTE, H.; ZIMMERMANN, K.; NIEDERSCHUH, S.; HELBIG, T.; VOGES, D.; HUSUNG, I.; VOLKOVA, T.; WILL, C.; BEHN, C., 2014. Technical, non-visual characterization of substrate contact using carpal vibrissae as a biological model: an overview. In: *Proceedings of the 58th IWK, Ilmenau Scientific Colloquium: shaping the future by engineering.* ilmedia, Available from urn:nbn:de:gbv:ilm1-2014iwk–175:8.

SCHOLZ, G. R.; RAHN, C. D., 2004. Profile sensing with an actuated whisker. *IEEE Transactions on Robotics and Automation.* Vol. 20, pp. 124–127. Available from DOI: 10.1109/TRA.2003.820864.

SCHROEDER, C. L.; HARTMANN, M. J. Z., 2012. Sensory prediction on a whiskered robot: a tactile analogy to „optical flow“. *Frontiers in Neurorobotics.* Vol. 6, pp. article 9. Available from DOI: 10.3389/fnbot.2012.00009.

SCHULTZ, A. E.; SOLOMON, J. H.; PESHKIN, M. A.; HARTMANN, M. J. Z., 2005. Multifunctional whisker arrays for distance detection, terrain mapping, and object feature extraction. In: *Proceedings of the 2005 IEEE International Conference on Robotics and Automation.* IEEE, pp. 2588–2593. Available from DOI: 10.1109/ROBOT.2005.1570503.

SCHWARZ, C., 2016. The slip hypothesis: tactile perception and its neuronal bases. *Trends in Neurosciences*. Vol. 39, pp. 449–462. Available from DOI: 10.1016/j.tins.2016.04.008.

SIMONS, D. J., 1995. Neuronal integration in the somatosensory whisker/barrel cortex. In: *The Barrel Cortex of Rodents*. Ed. by JONES, E. G.; DIAMOND, I. T. Boston, MA: Springer, pp. 263–297. Available from DOI: 10.1007/978-1-4757-9616-2_6.

SIMONY, E.; BAGDASARIAN, K.; HERFST, L.; BRECHT, M.; AHISSAR, E.; GOLOMB, D., 2010. Temporal and spatial characteristics of vibrissa responses to motor commands. *The Journal of Neuroscience*. Vol. 30, pp. 8935–8952. Available from DOI: 10.1523/JNEUROSCI.0172-10.2010.

SINGH, P.; GOSS, V. G. A., 2019. The clamped-free rod under inclined end forces and transitions between equilibrium configurations. *Journal of Engineering Mathematics*. Vol. 59, pp. 14. Available from DOI: 10.1007/s10665-019-10009-7.

SKERRA, B., 2000. *Handbuch Molchtechnik*. Essen: Vulkan-Verlag.

SODERQUIST, D. R. R., 2002. *Sensory Processes*. Thousand Oaks, CA: SAGE Publications. ISBN 978-0761923336.

SOFRONIEW, N. J.; SVOBODA, K., 2015. Whisking. *Current Biology*. Vol. 25, pp. PR137–R140. Available from DOI: 10.1016/j.cub.2015.01.008.

SOFRONIEW, N. J.; VLASOV, Y. A.; HIRES, S. A.; FREEMAN, J.; SVOBODA, K., 2015. Neural coding in barrel cortex during whisker-guided locomotion. *eLife*. Vol. 4, pp. e12559. Available from DOI: 10.7554/eLife.12559.

SOLOMON, J. H.; HARTMANN, M. J. Z., 2006. Robotic whiskers used to sense features. *Nature*. Vol. 443, pp. 525. Available from DOI: 10.1038/443525a.

SOLOMON, J. H.; HARTMANN, M. J. Z., 2011. Radial distance determination in the rat vibrissal system and the effects of Weber's law. *Philosophical Transactions of the Royal Society B: Biological Sciences*. Vol. 366, pp. 3049–3057. Available from DOI: 10.1098/rstb.2011.0166.

SONG, P.; YU, Y.; ZHANG, X., 2017. Impedance control of robots: an overview. In: *2017 2nd International Conference on Cybernetics, Robotics and Control*. IEEE, pp. 51–55. Available from DOI: 10.1109/CRC.2017.20.

STAROSTIN, E. L.; GRANT, R. A.; DOUGILL, G.; VAN DER HEIJDEN, G. H. M.; GOSS, V. G. A., 2020. The Euler spiral of rat whiskers. *Science Advances*. Vol. 6, pp. eaax5145. Available from DOI: 10.1126/sciadv.aax5145.

STEIGENBERGER, J., 2013. A continuum model of passive vibrissae. In: *Preprint / Technische Universität Ilmenau, Institut für Mathematik*. Ilmenau, Germany: Institut für Mathematik. No. M 13/03.

STEIGENBERGER, J.; BEHN, C.; WILL, C., 2015. Mathematical model of vibrissae for surface texture detection. In: *Preprint / Technische Universität Ilmenau, Institut für Mathematik*. Ilmenau, Germany: Institut für Mathematik. No. M 15/03.

STUBBIG, L.; LICHTENHELDT, R.; BECKER, F.; ZIMMERMANN, K., 2017. Model-based development of a compliant locomotion system for a small scout rover. In: *Proceedings of the 59th IWK, Ilmenau Scientific Colloquium: Engineering for a Changing World*. ilmedia, Available from urn:nbn:de:gbv:ilm1-2017iwk–012:0.

STÜTTGEN, M. C.; RÜTER, J.; SCHWARZ, C., 2006. Two psychophysical channels of whisker deflection in rats align with two neuronal classes of primary afferents. *The Journal of Neuroscience*. Vol. 26, pp. 7933–7941. Available from DOI: 10.1523/JNEUROSCI.1864-06.2006.

SZABÓ, I., 1963. *Einführung in die Technische Mechanik*. Berlin, Heidelberg: Springer.

SZWED, M.; BAGDASARIAN, K.; AHISSAR, E., 2003. Encoding of vibrissal active touch. *Neuron*. Vol. 40, pp. 621–630. Available from DOI: 10.1016/S0896-6273(03)00671-8.

TOWAL, R. B.; HARTMANN, M. J. Z., 2006. Right-left asymmetries in the whisking behavior of rats anticipate head movements. *The Journal of Neuroscience*. Vol. 26, pp. 8838–8846. Available from DOI: 10.1523/JNEUROSCI.0581-06.2006.

TOWAL, R. B.; QUIST, B. W.; GOPAL, V.; SOLOMON, J. H.; HARTMANN, M. J. Z., 2011. The morphology of the rat vibrissal array: a model for quantifying spatiotemporal patterns of whisker-object contact. *PLoS Computational Biology*. Vol. 7, pp. e1001120. Available from DOI: 10.1371/journal.pcbi.1001120.

TOWAL, R. B.; QUIST, B. W.; SOLOMON, J. H.; HARTMANN, M. J. Z., 2012. Active sensing: head and vibrissal velocity during exploratory behaviors of the rat. In: *Frontiers in Sensing*. Ed. by BARTH, F. G.; HUMPHREY, J. A. C.; SRINIVASAN, M. V. Vienna: Springer, pp. 209–224. Available from DOI: 10.1007/978-3-211-99749-9_14.

VAXENBURG, R.; WYCHE, I.; SVOBODA, K.; EFROS, A. L.; HIRES, S. A., 2018. Dynamic cues for whisker-based object localization: an analytical solution to vibration during active whisker touch. *PLoS Computational Biology*. Vol. 14, pp. e1006032. Available from DOI: 10.1371/journal.pcbi.1006032.

VAZIRI, A.; JENKS, R. A.; BOLOORI, A.; STANLEY, G. B., 2007. Flexible probes for characterizing surface topology: from biology to technology. *Experimental Mechanics*. Vol. 47, pp. 417–425. Available from DOI: 10.1007/s11340-007-9046-8.

VINCENT, S. B., 1912. *The function of the vibrissae in the behavior of the white rat*. PhD thesis. University of Chicago: Department of Psychology.

VINCENT, S. B., 1913. The tactile hair of the white rat. *The Journal of Comparative Neurology*. Vol. 23, pp. 1–34. Available from DOI: 10.1002/cne.900230101.

VOGEL, S., 1998. *Cats' paws and catapults: Mechanical worlds of nature and people*. New York, NY: W. W. Norton & Company, Inc. ISBN 0-393-04641-9.

VOGES, D.; CARL, K.; KLAUER, G. J.; UHLIG, R.; SCHILLING, C.; BEHN, C.; WITTE, H., 2012. Structural characterization of the whisker system of the rat. *IEEE Sensors Journal*. Vol. 12, pp. 332–339. Available from DOI: 10.1109/JSEN.2011.2161464.

VOLK, R., 2013. *Rauheitsmessung: Theorie und Praxis*. Berlin: Beuth Verlag. ISBN 978-3-410-23882-9.

VOLKOVA, T.; ZEIDIS, I.; WITTE, H.; SCHMIDT, M.; ZIMMERMANN, K., 2016. Analysis of the vibrissa parametric resonance causing a signal amplification during whisking behaviour. *Journal of Bionic Engineering*. Vol. 13, pp. 312–323. Available from DOI: 10.1016/S1672-6529(16)60304-9.

WEAVER, W.; TIMOSHENKO, S. P.; YOUNG, D. H., 1990. *Vibration problems in engineering*. New York: John Wiley & Sons.

WEGIRIYA, H.; HERZIG, N.; ABAD, S.-A.; SADATI, S. M. H.; NANAYAKKARA, T., 2019. A stiffness controllable multimodal whisker sensor follicle for texture comparison. *IEEE Sensors Journal*. Pp. 1–1. Available from DOI: 10.1109/JSEN.2019.2951755.

WIERTLEWSKI, M., 2013. *Reproduction of tactual textures: transducers, mechanics and signal encoding*. London: Springer. ISBN 978-1-4471-4840-1.

WILL, C., 2018. *Continuum models for biologically inspired tactile sensors : theory, numerics and experiments*. PhD thesis. Technische Universität Ilmenau.

WILL, C.; BEHN, C.; STEIGENBERGER, J., 2017. Object contour scanning using elastically supported technical vibrissae. *Journal of Applied Mathematics and Mechanics*. Vol. 79, pp. 1–17. Available from DOI: 10.1002/zamm.201600161.

WILLIAMS, C. M.; KRAMER, E. M., 2010. The advantages of a tapered whisker. *PLoS ONE*. Vol. 5, pp. e8806. Available from DOI: 10.1371/journal.pone.0008806.

WITTE, H.; FREMEREY, M.; WEYRICH, S.; MÄMPEL, J.; FISCHHEITER, L.; VOGES, D.; ZIMMERMANN, K.; SCHILLING, C., 2013. Biomechatronics is not just biomimetics. In: *2013 9th International Workshop on Robot Motion and Control*. IEEE, pp. 74–79. Available from DOI: 10.1109/RoMoCo.2013.6614587.

WITTE, H.; LUTHERDT, S.; SCHILLING, C., 2004. Biomechatronics: how much biology does the engineer need? In: *Proceedings of the 2004 IEEE International Conference on Control Applications*. IEEE, pp. 944–948. Available from DOI: 10.1109/CCA.2004.1387491.

WITTE, H.; SCHILLING, C., 2017. The concept of biomechatronic systems as a means to support the development of biosensors. *International Journal of Biosensors & Bioelectronics*. Vol. 2, pp. 114–115. Available from DOI: 10.15406/ijbsbe.2017.02.00030.

WOLFE, J.; HILL, D. N.; PAHLAVAN, S.; DREW, P. J.; KLEINFELD, D.; FELDMAN, D. E., 2008. Texture coding in the rat whisker system: slip-stick versus differential resonance. *PLoS Biology*. Vol. 6, pp. e215. Available from DOI: 10.1371/journal.pbio.0060215.

YAN, S. W. Y.; GRAFF, M. M.; HARTMANN, M. J. Z., 2016. Mechanical responses of rat vibrissae to airflow. *Journal of Experimental Biology*. Vol. 219, pp. 937–948. Available from DOI: 10.1242/jeb.126896.

YANG, A. E.-T.; BELLI, H.; HARTMANN, M. J. Z., 2019. Quantification of vibrissal mechanical properties across the rat mystacial pad. *Journal of Neurophysiology*. Vol. 121, pp. 1879–1895. Available from DOI: 10.1152/jn.00869.2016.

YANLI, B.; WEI, Z.; YANCHUN, X.; JUN, Z.; XIAOMING, T., 1998. Relationship between structure and function of mammalian vibrissa. *Journal of Forestry Research*. Vol. 9, pp. 273–282. Available from DOI: 10.1007/BF02912334.

ZENTNER, L.; GRIEBEL, S.; HÜGL, S., 2019. Fluid-mechanical compliant actuator for the insertion of a cochlear implant electrode carrier. *Mechanism and Machine Theory*. Vol. 142, pp. 16. Available from DOI: 10.1016/j.mechmachtheory.2019.103590.

ZUO, Y.; PERKON, I.; DIAMOND, M. E., 2011. Whisking and whisker kinematics during a texture classification task. *Philosophical Transactions of the Royal Society B: Biological Sciences*. Vol. 366, pp. 3058–3069. Available from DOI: 10.1098/rstb.2011.0161.

Standards

Deutsches Institut für Normung e.V., 2011. DIN 18196: Earthworks and foundations - Soil classification for civil engineering purposes. Berlin: Beuth Verlag, 2011-05-00.

Deutsches Institut für Normung e.V., 1982. DIN 4760: Gestaltabweichung: Begriffe, Ordnungssystem. Berlin: Beuth Verlag, 1982-06-00.

Deutsches Institut für Normung e.V., 1993. DIN 50323-3: Tribology; friction; definitions, types, conditions, quantities (withdrawn). Berlin: Beuth Verlag, 1993-12-00.

Deutsches Institut für Normung e.V., 2011. EN 10270-1: Steel wire for mechanical springs - Part 1: Patented cold drawn unalloyed spring steel wire + A1:2017. Berlin: Beuth Verlag, 2017-09-00.

Bureau International des Poids et Mesures, 2008. ISO/IEC GUIDE 98-3: Uncertainty of measurement - Part 3: Guide to the expression of uncertainty in measurement (GUM:1995). Paris: Bureau International des Poids et Mesures, 2008(E).

Deutsches Institut für Normung e.V., 2015. ISO 18458: Biomimetics - Terminology, concepts and methodology. Berlin: Beuth Verlag, 2016-08-00.

Deutsches Institut für Normung e.V., 1996. ISO 3274: Geometrical Product Specifications (GPS) - Surface texture: Profile method - Nominal characteristics of contact (stylus) Instruments. Berlin: Beuth Verlag, 1998-04-00.

Deutsches Institut für Normung e.V., 2010. ISO 4287: Geometrical product specifications (GPS) - surface texture: profile method - terms, definitions and surface texture parameters. Berlin: Beuth Verlag, 2010-07-00.

Deutsches Institut für Normung e.V., 2000. ISO 6344-2: Coated abrasives - Grain size analysis - Part 2: Determination of grain size distribution of macrogrits P12 to P220. Berlin: Beuth Verlag, 2000-04-00.

Deutsches Institut für Normung e.V., 2017. ISO 6344-3: Coated abrasives - Grain size analysis - Part 3: Determination of grain size distribution of microgrits P240 to P2500. Berlin: Beuth Verlag, 2017-04-00.

Deutsches Institut für Normung e.V., 1999. ISO 8785: Geometrical product specifications (GPS) - surface imperfections - terms, definitions and parameters. Berlin: Beuth Verlag, 1999-10-00.

VDI/VDE-Gesellschaft Mess- und Automatisierungstechnik, 2018. VDI/VDE-Richtlinien 2617: Accuracy of coordinate measuring machines characteristics and their testing form measurement with coordinate measuring machines. Berlin: Beuth Verlag, 2018-07-00.

VDI/VDE-Gesellschaft Mess- und Automatisierungstechnik, 2016. VDI/VDE-Richtlinien 2631: Form measurement principles. Berlin: Beuth Verlag, 2016-08-00.

A | Comparison of theoretical models and pre-analysis

The experiments in Chapter 5 are accompanied by simulations to enhance the understanding of the results. The finite element software ANSYS (see Table B.1) is used to design the simulation in a sufficiently detailed manner, see Section A.2.

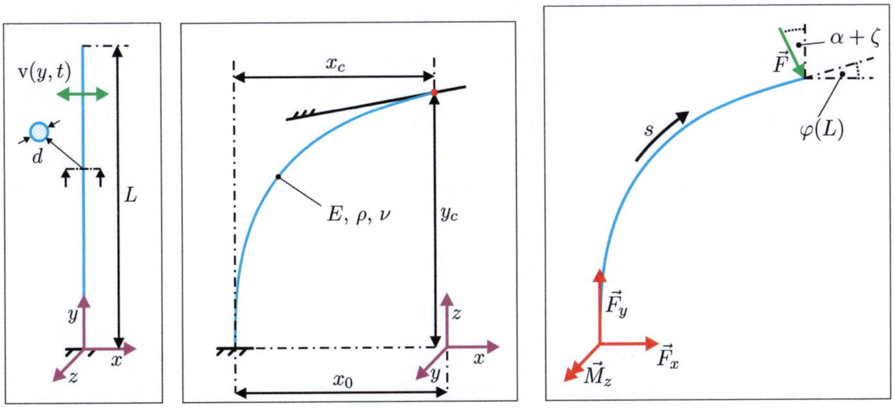

(a) (b) (c)

Figure A.1 – **(a)**- Unloaded beam: the geometrical dimensions of the straight, slender, cylindrical beam are given by length L and diameter d. The displacement of a cross section of the beam duirng free vibrtaion is described by the function $\mathrm{v}(y, t)$. **(b)**- Loaded beam: the material is characterized by Young's modulus E, density ρ and Poisson's ratio ν. The current contact location is indicated by the coordinates $(x_c\,, y_c)$ and depicted as red dot. **(c)**- Free body diagram: the elastica of the deformed beam is described using natural coordinates in the x-y-plane with arc length $s \in [0,\, L]$. The contact can occur along the beam including tip as well as a tangential contact. If $s = s_c = L$ and $\varphi(L) = \varphi_c \neq \alpha$ there is tip contact, else if $s = s_c \leq L$ and $\varphi(s_c) = \varphi_c = \alpha$ there is tangential contact. The orientation of the contact force $\vec{F}$ is also influenced by the friction angle ζ. The support reactions are defined as: $\vec{F}_x$, $\vec{F}_y$ as clamping forces and $\vec{M}_z$ as clamping bending moment.

The simulations are done in ANSYS with the goal to assess the experimental data. They do not aim to develop *the perfect model* of a vibrissa which seems to be more a self-purpose than a useful contribution because theoretical models are always constrained and of limited validity compared to reality. Therefore, the model within the limits of ANSYS is sufficiently accurate. The finite element simulation is verified by a simulation with respect to the Euler-Bernoulli beam theory in MATLAB (see Table B.1), see Section A.1. This verification is a spin-off of the work [Scharff et al.; 2018] and in parts reused as well as extended, see Section A.3.

A.1 Euler-Bernoulli beam model and reconstruction algorithm

The original procedure is presented in [Steigenberger; 2013] and the authors of [Scharff et al.; 2017b], [Merker et al.; 2018] extended it. The following assumptions and considerations are done, compare Figure A.1:

- The vibrissa hair/sensor shaft is assumed as an Euler-Bernoulli beam undergoing large deflection due to bending deformation;
- the FSC/sensor support is modeled as clamping, which obviously presents strong simplification of the natural paragon, see Section 4.2;
- the displacement/rotation of the clamping and the displacement of the cross-sections due to bending deformation of the beam are quasi-statically and restricted to the x-y-plane (support reactions: F_x, F_y and M_z);
- gravity is neglected;
- the beam consists of a linear elastic Hooke's material with Young's modulus E, density ρ and Poisson's ratio ν;
- the beam has a circular cross section with a diameter d and a slender, straight body with a length L;
- the tapered shape and inherent curvature of a natural vibrissa are neglected and, consequently, the taper ratio is $\vartheta = 1$ and the inherent curvature $\kappa_0 = 0$;
- the contact between beam and object is a frictional point contact within the limits of dry friction, see Subsection 2.2.2;
- the object is rigid and parameterized by its slope α.

These restrictions and assumptions are considered to describe the deformation of the beam. More detailed versions of this model are reported in [Scharff et al.; 2016],

[Scharff et al.; 2017a], [Scharff et al.; 2017c]. The starting point to describe the elastica of the deformed beam is the presentation of a plane curve in natural coordinates parameterized by its slope angle $\varphi(s)$ depending on the arc length s, see (A.1):

$$\left.\begin{aligned} \frac{\mathrm{d}x(s)}{\mathrm{d}s} &= \cos(\varphi(s)) \\ \frac{\mathrm{d}y(s)}{\mathrm{d}s} &= \sin(\varphi(s)) \\ \frac{\mathrm{d}\varphi(s)}{\mathrm{d}s} &= \kappa(s) \end{aligned}\right\} \tag{A.1}$$

Here, $\kappa(s)$ represents the curvature of the beam. The connection between the geometrical description and the mechanical behavior is given by the constitutive law of Euler, e.g., [Szabó; 1963], see (A.2):

$$\kappa(s) = \frac{M_{bz}(s)}{E \cdot I_z} \tag{A.2}$$

The Young's modulus E, as well as the second moment of area I_z, is given by the structure of the beam. The bending moment $M_{bz}(s)$ along the beam results out of the cross-product of contact position and contact force applying the method of sections:

$$M_{bz}(s) = \begin{cases} F\left[\sin(\alpha+\zeta)\,(y(s)-y_c) + \cos(\alpha+\zeta)\,(x(s)-x_c)\right] & s \in (0, s_c) \\ 0 & s \in (s_c, L) \end{cases} \tag{A.3}$$

with (x_c, y_c) as location of the contact point along the elastica, considering that $s_c \in (0, L]$. Combining (A.1)-(A.3), a non-linear system of ordinary differential equation of first order results:

$$\left.\begin{aligned} \frac{\mathrm{d}x(s)}{\mathrm{d}s} &= \cos(\varphi(s)) \\ \frac{\mathrm{d}y(s)}{\mathrm{d}s} &= \sin(\varphi(s)) \\ \frac{\mathrm{d}\varphi(s)}{\mathrm{d}s} &= \frac{M_{bz}(s)}{E \cdot I_z} \end{aligned}\right\} \tag{A.4}$$

with the belonging boundary conditions:

$$\left.\begin{aligned} x(0) &= x_0; & x(s_c) &= x_c; \\ y(0) &= 0\,; & y(s_c) &= y_c; \\ \varphi(0) &= \varphi_0; & \varphi(s_c) &= \varphi_c; \end{aligned}\right\} \tag{A.5}$$

Changing the value of x_0 or φ_0 in (A.5) the quasi static displacement of the clamping is realized. Hence, these variables can be interpreted as a kinematic drive of the system. For each increment (A.4) is solved numerically. The initial situation is that the beam contacts the surface, the resulting contact force has components in y-direction only.

Both, a tip contact or a tangential contact along the beam are possible. Starting a displacement or rotation of the clamping by incrementally changing x_0 or φ_0, an x-component of the contact force arises as a consequence of friction. The direction of the x-component of the contact force is orientated in opposite direction of the displacement. The contact point sticks to the surface until the μ_s is reached.

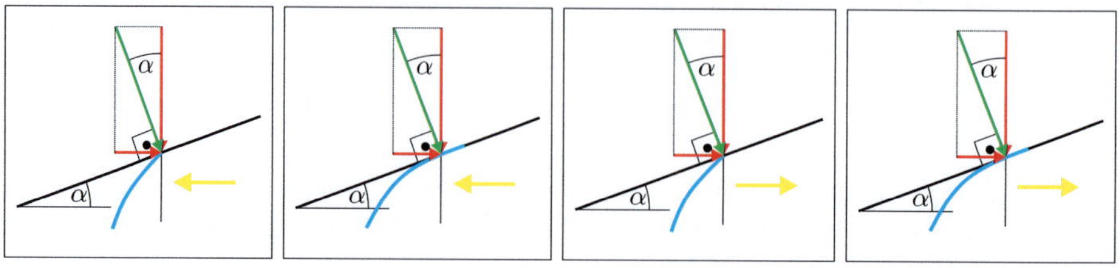

(a) tip contact **(b)** tangential contact **(c)** tip contact **(d)** tangential contact

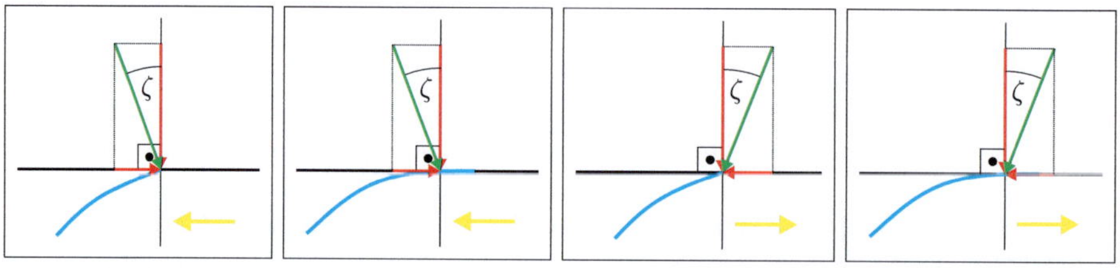

(e) tip contact **(f)** tangential contact **(g)** tip contact **(h)** tangential contact

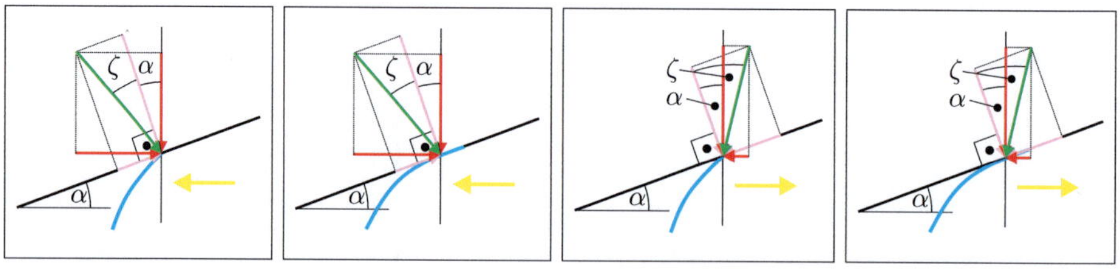

(i) tip contact **(j)** tangential contact **(k)** tip contact **(l)** tangential contact

Figure A.2 – Illustration of different contact scenarios during scanning an object with a straight curve as shape: green - resulting contact force; red - components of the contact force in the global frame; pink - normal and tangential components of the contact force with respect to α; yellow - displacement direction of the support; cyan - section of the deformed sensor shaft. In **(a)** - **(d)** $\zeta = 0$, $\alpha \neq 0$; **(e)** - **(h)** $\zeta \neq 0$, $\alpha = 0$ and **(i)** - **(l)** $\zeta \neq 0$, $\alpha \neq 0$.

Afterward, the contact point starts sliding. Here, the support remains in its current position and gets not further displaced while the contact point quasi-statically displaces along the object shape. The displacement of the contact point is realized by an incremental change of x_c in (A.5). When the present μ is equal to μ_d, the contact point sticks again, and the described procedure repeats. The present μ can be determined by evaluating F_t and F_n during a scanning sweep (A.6).

$$\tan(\zeta) = \frac{|F_t|}{|F_n|} = \mu \quad ; \quad \mu_d \leq \mu \leq \mu_s \tag{A.6}$$

Since the object shape is parameterized by α, the present μ cannot be determined directly using the captured support reactions. The slope angle of the object α must be taken into account. The friction angle ζ, slope angle α, and support reactions F_x and F_y are related as follows:

$$\alpha - \zeta \cdot \mathrm{sgn}(\Delta x_0) = -\arctan\left(\frac{F_x}{F_y}\right) \tag{A.7}$$

All possible contact situations are depicted in Figure A.2 whereby all situations apply for sticking and sliding of the contact point and $\alpha \in [-\frac{\pi}{2}, \frac{\pi}{2}]$. An a priori knowledge of α or ζ in every situation is necessary to solve (A.4). This condition can be only guaranteed if α or ζ are constant. This is the case if friction is neglected with ζ equal zero, Figure A.2(a)-(d) else if the object shape is represented by a straight curve with a constant α, Figure A.2(e)-(l). A further difference results out of the displacement direction since the direction of $\vec{F}$ is related to the direction of the displacement.

Algorithm for contact point location detection

The following approach is taken from [Scholz, Rahn; 2004]. The assumptions regarding the Euler-Bernoulli beam model in Section A.1 are valid to apply that algorithm to determine the position of the contact point. For the determination of the contact point, (A.4) must be solved whereby the bending moment is a function of s given by (A.8).

$$M_{bz}(s) = \begin{cases} M_z - F_y \cdot x(s) + F_x \cdot y(s) & s \in (0, s_c) \\ 0 & s \in (s_c, L) \end{cases} \tag{A.8}$$

The solution is obtained by numerical integration of (A.4) using a Runge-Kutta-Fehlberg method basically established in the MATLAB function `ode45`$(\cdot)$ considering that $s_c \in (0, L]$. The integration stops if the bending moment (A.8) tends to zero because then the contact point position is reached.

A.2 Finite element analysis

The finite element analysis is done using the software ANSYS, more specific, Mechanical APDL, see Table B.1. Similar to Section A.1, the sensor shaft is modeled as a beam and the object is assumed to be rigid. The beam is characterized by the 3D- beam element `beam188` with a linear shape function. The element basis is the Timoshenko beam theory. Therefore, it includes shear deformation in contrast to the Euler-Bernoulli beam theory. The cross section is modeled to be rigid that means there is no axial elongation of the beam and, consequently, no change of the inertia properties of the cross section. The clamping is realized by constraining all degrees of freedom of the first node located at the boundary. The beam is meshed by 100 elements of constant size. Every element of type `beam188` is superimposed by a contact element `conta177`. Since, in the present work the scanning process is limited to the x-y-plane (see Chapter 4), the object shape can be modeled as a curve. Straight object contours like in, e.g., Sections 5.3 or A.3 are modeled by one single straight line using `tshap` function. Arbitrary object shapes as analyzed in Section 5.1 are generated by an adequate polynomial approximation. In either case, the object is meshed by `targe170` elements. The combination of `conta177` and `targe170` results in a line to surface/line contact which is limited to two dimensions done by previous assumptions. The contact is defined to be forced based and considers dry friction whereby the transition from sticking to sliding depends on an exponential decay. This combination of elements is capable to cover single as well as multiple contacts between surface and beam. The impact of the beam on the object surface is damped by a coefficient of restitution of cr = 2e-8. A small coefficient of restitution dissipates energy of the system, and consequently, the beam does not snap on and off the object. This is confirmed by observations regarding natural vibrissae [Hartmann et al.; 2003] and by observations during the experiments of the present work. Structural and material damping are considered to be combined in the stiffness matrix multiplier β = 2.15e-5 within the scope of Rayleigh damping. The value is chosen based on the findings in Subsection B.4.2, in a way that the damping ratio in simulation is equal to the experimental one. As a contact algorithm, an `augmented Lagrangian method` procedure is chosen. Here, the penetration between beam and object is forced to reach defined penetration tolerance. Therefore, a sequence of penalty methods with respect to different, program determined, contact normal stiffnesses is applied. During the simulation, the contact stiffness is ongoingly redefined by the program in order to guarantee adequate contact conditions. Quasi-static, full transient, and modal analyses are used in the present work. The quasi-static analysis neglects all dynamic effects, which are included in the transient analysis. Both analyses consider large deflections. The simulations resolve 1000 time steps per second. Thus, high frequencies can be detected by evaluating the simulated support reactions. The modal analysis is combined with a static analysis in order to pre-stress the beam and determine the corresponding natural frequencies.

A.3 Verification of the theoretical models

The model based on Euler-Bernoulli beam theory is used to verify the model incorporated in ANSYS, e.g., to demonstrate that 100 beam elements are sufficient to discretize the beam with respect to the deformation (see Subsections A.3.1 und A.3.2) and dynamical behavior (see Subsection A.3.3). The geometrical and material properties are given in Table B.3. For the verification a straight object contour without inclination is used and placed $\eta = 0.07\,\mathrm{m}$ away[3] from the clamping. The dynamic friction coefficient is $\mu_d = 0.2$ and the static one is $\mu_s = 0.24$, these values are conform with a steel-to-steel contact pair.

A.3.1 Translational actuation

To solve (A.4) for a translational displacement (see Figure A.3) the boundary conditions (A.5) are specified as:

$$\left.\begin{array}{rlrcl} x(0) = & x_0; & x(s_c) & = & x_c; \\ y(0) = & 0\ ; & y(s_c) & = & y_c; \\ \varphi(0) = & \dfrac{\pi}{2}; & \varphi(s_c) & = & \varphi_c; \end{array}\right\} \tag{A.9}$$

with

$$x_0 = 0\,\mathrm{m}\ (-\Delta x_0)\ -0.01\,\mathrm{m} \tag{A.10}$$

In the case of the finite element analysis, the translational displacement is realized by manipulating the constraint of the clamping in x-direction, considering a constant shifting velocity $v = 0.001\,\mathrm{m\,s^{-1}}$. Figure A.3 shows a selection of elasticae of the beam during protraction. The contact is always between tip and object, compare Figure A.2 (e) & (g). The corresponding support reactions, for pro- and retraction are given in Figure A.4. The illustrated results are limited to smaller intervals because of the visibility of the signal characteristics. The constant distance between clamping and object surface η causes a constant stationary signal part for all support reactions in Figure A.4. If there is no friction and dynamic effects are neglected, F_x is zero and F_y and M_z are constant and there is no notable difference between pro- and retraction (green curves).

[3]Remind the length of the sensor shaft is $L = 0.1\,\mathrm{m}$

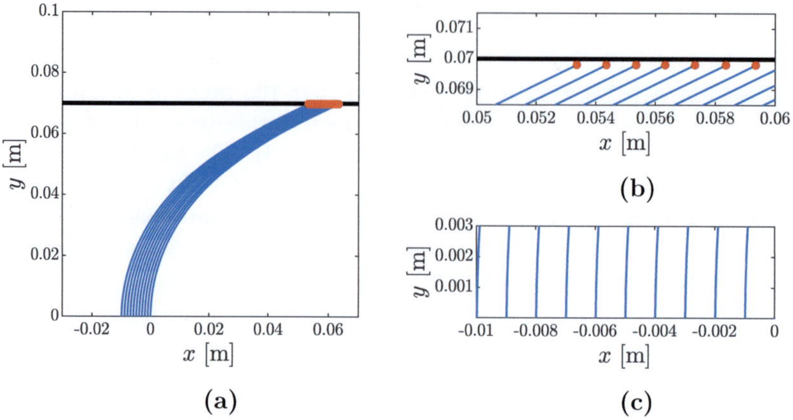

Figure A.3 – **(a)** Elasticae of the beam during scanning a straight, rough object with a translational displacement of the support. The contact point is highlighted in red. The contact point is highlighted in red. In **(b)** the tip of the beam is in focus and in **(c)** the clamping position.

Including dynamic effects, there are very small oscillations superimposing the stationary signal (magenta curves). Compared to the previous case, the signals are very similar besides the small oscillations. Also including dynamic effects, the signals for pro- and retraction are alike. Extending the simulation to a frictional contact behavior, but neglecting dynamical effects, the signal characteristics change (green curve). At the beginning of the scanning sweep, there is a prominent signal change followed by an abrupt signal drop. The signal change corresponds to sticking of the beam tip on the object surface. Therefore, a static friction force is present until μ_s is reached, then the tip starts sliding until the end of the scanning sweep. Compared to the frictionless simulations, F_x is not zero and its orientation is opposite to the displacement direction resulting in a sign change for protraction (negative) and retraction (positive), see Figures A.4(a)-(c) and (d)-(e). The support reaction F_y decreases and M_z increases in the case of protraction and a reversed behavior for retraction. This behavior is caused by the influence of the friction force. In the case of protraction, the bending moment increases resulting in a stronger deformation of the beam and, consequently, in a less load in y-direction. In contrast, while retraction, the load in y-direction gets forced. The red curves in Figure A.4 correspond to the solution of (A.4). The assumptions are equal to the ones regarding the green curves, dynamical effects are neglected and friction is included. Both simulations show very similar results across all support reactions. They are nearly identical until the first edge and the following drop. Afterward, the signals belonging to (A.4) are characterized by a continuous sequence of edges and drops, indicating a stick-slip motion of the contact point.

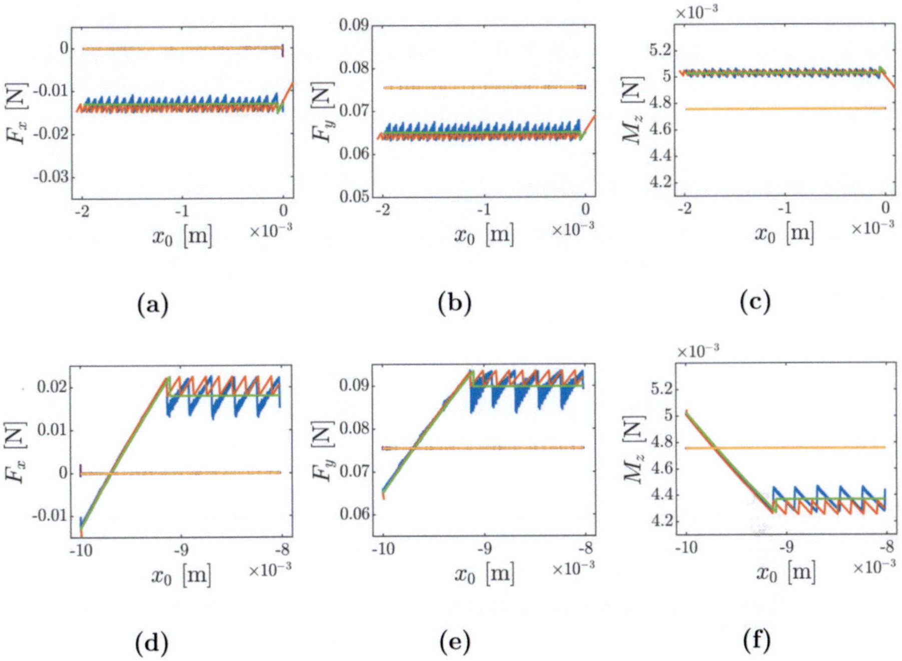

Figure A.4 – The first row shows the support reactions for a translational protraction phase und the second row for a retraction one with: **(a)**&**(d)**- F_x; **(b)**&**(e)**- F_y; **(c)**&**(f)**- M_z. Blue curves correspond to a full transient simulation considering friction; green curves to a quasi-static simlation considering friction; magenta curves to a full transient simulation without friction; yellow curves to a quasi-static simulation without friction and red curves to the soultion of (A.4) neglecting dynamical effects but including friction.

Whereby, the lower limit of the edge corresponds to sliding friction force and the upper one to the sticking friction force. The sawtooth-like signals superimpose stationary signals that follow the same behavior as described for the green curves. Comparing pro- and retraction, Figures A.4(a)-(c) to (d)-(e), the amplitudes and the frequency of the stick-slip motion change. In the case of protraction, the stick-slip motion has a high frequency and small amplitude, whereby for retraction the frequency is low and the amplitude large. Further changes of the signals can be observed by including dynamical effects as well as friction in the simulation (blue curves). All support reactions hint at a stick-slip motion. The first edge aligns to the edges of the green and red curves. The following stick-slip oscillation has remarkably larger amplitudes than the red signals, but besides this, they exhibit alike characteristics. Furthermore, the support reactions include other high-frequency components.

These additional signal components seem to influence more F_y than F_x and M_z because the signals are *noisy*. The lower limit of the amplitude of the stick-slip motion is smaller than for the simulations belonging to the red curve. This change causes a difference of the frequency since the sticking period gets longer.

A.3.2 Rotatory actuation

For the rotation of the sensor shaft, it is defined that the support reactions remain in the initial coordinate system. The corresponding boundary conditions are given in (A.11):

$$\left.\begin{array}{rlrcl} x(0) = & 0\,; & x(s_c) & = & x_c; \\ y(0) = & 0\,; & y(s_c) & = & y_c; \\ \varphi(0) = & \varphi_0; & \varphi(s_c) & = & \varphi_c; \end{array}\right\} \tag{A.11}$$

with

$$\varphi_0 = 0.27\,\pi\;(\Delta\varphi_0)\;0.33\,\pi \tag{A.12}$$

as actuation sequence. The interval for φ_0 is chosen to generate a scanning sweep that is alike to the one for a translational displacement in Subsection A.3.1. In contrast to the translational displacement, for the rotation, only the protraction is analyzed. Retraction is not further evaluated because in the present configuration a retraction of the sensor shaft leads to buckling. In literature, it is reported that a displacement of a sensor shaft in the direction to the concave side of the sensor shaft can cause buckling with respect to the second buckling mode [Hires et al.; 2013], [Steigenberger et al.; 2015].

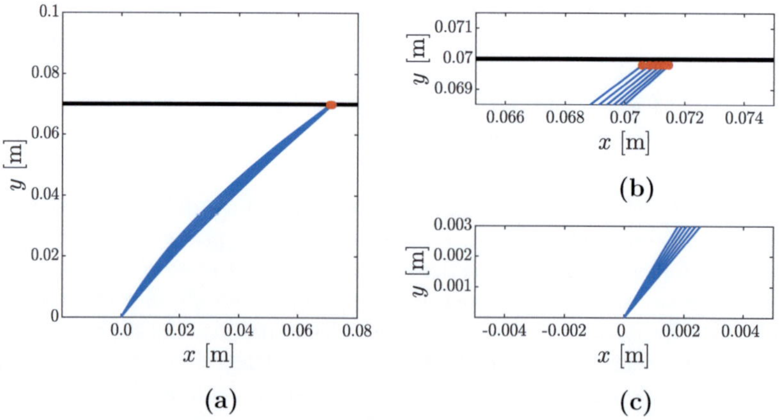

Figure A.5 – **(a)** Elasticae of the beam during scanning a straight, rough object with a rotation of the support. The contact point is highlighted in red. In **(b)** the tip of the beam is in focus and in **(c)** the clamping position.

Finally, this effect can result in a so-called turnover of the sensor shaft [Scharff et al.; 2017b]. This effect is not in the focus of the present section and, consequently, retraction is not further analyzed as already mentioned. In Figure A.5, the displacement and deformation of the beam are shown. But, the displacement of the tip across the object surface is small. Furthermore, there are only contacts between the tip of the beam and the object surface because the distance between clamping and object surface $\eta > 0.457 \cdot L$ which is found to be the limit for the presence of a tangential contact scenario [Steigenberger; 2013]. The support reactions for all types of simulation are illustrated in Figure A.6. Again, if friction, as well as dynamical effects, are neglected, there is no force in x-direction (yellow curve), see Figure A.6(a). The support reactions F_y and M_z are constantly increasing for an increasing φ_0. A similar behavior can be observed if dynamic effects are included (magenta curve). Here, there is additionally a small oscillating signal component superimposing the stationary signal part. Compared to the simulations due to a translational displacement in Subsection A.3.1, the amplitude of this overlaying oscillation is larger in observing the force F_x, see Figure A.6(a). There is no remarkable change for F_y and M_z. Including frictional effects and neglecting dynamical ones, F_x exhibits a constantly changing trend (yellow curve/ANSYS, red curve/solution of (A.4)). Both signals show a high-frequency sawtooth-like trend whereby the frequencies are too high to be clearly visible in Figure A.6(b) and (c). Furthermore, the yellow curve shows regularly spaced peaks and not regular appearing intervals with small amplitudes. The peaks are best visible for F_x. The full transient simulation also shows these notable peaks and a high-frequency sawtooth-like signal with a constantly changing stationary part.

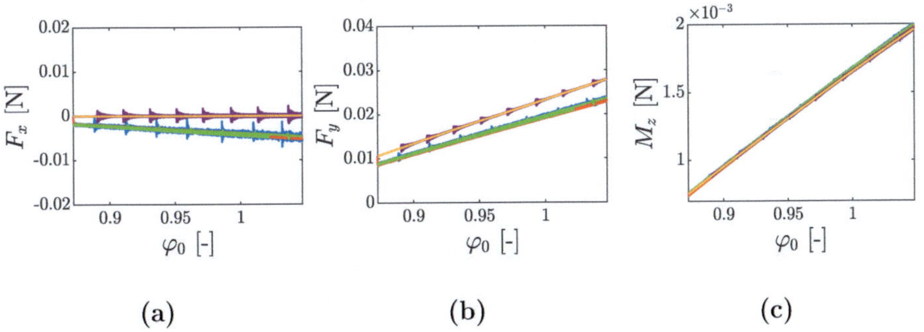

Figure A.6 – The support reactions for a rotatory protraction phase are illustrated in: **(a)**- F_x; **(b)**- F_y; **(c)**- M_z. Blue curves correspond to a full transient simulation considering friction; green curves to a quasi-static simlation considering friction; magenta curves to a full transient simulation without friction; yellow curves to a quasi-static simulation without friction and red curves to the soultion of (A.4) neglecting dynamical effects but including friction.

Again, the peaks are best visible in the case of F_x. But, they are not aligned to the peaks of the yellow curve, they seem to be shifted. Comparing simulation considering friction with the ones neglecting friction, F_x is oscillating around zero and there is a variable offset between the corresponding curves. For all four simulations, M_z is very similar.

The simulations with a rotating sensor require a much smaller numerical step size than needed for the translational simulations because of the higher stick-slip frequency, indicated by the sawtooth-like signals. So, for a constant η, a rotatory scanning sweep must be performed with a smaller time step size in simulation and a higher sampling frequency in experiment. In both cases, the amount of recorded data is larger than for a translational scanning sweep.

The different levels of simulations are consistent and are in accordance with the findings in Subsection 2.2.2 regarding friction and the Stick-Slip effect. In the next step, also the dynamic characteristics of the finite element model get verified.

A.3.3 Dynamic characteristics

The natural frequencies of the beam are calculated using again Euler-Bernoulli beam theory (A.13) to verify the results of the modal analysis done in ANSYS. The procedure follows [Weaver et al.; 1990], see also Figure A.1.

$$\frac{\partial^2 \mathrm{v}(y,t)}{\partial t^2} + \frac{4\,E\,I_z}{\rho\,\pi\,d^2} \cdot \frac{\partial^4 \mathrm{v}(y,t)}{\partial y^4} = 0 \tag{A.13}$$

The solution of (A.13) can be found by separation of variables whereby $v(x,t)$ is the deflection of the beam (A.14):

$$\mathrm{v}(y,t) = T(t) \cdot Y(y) \qquad \forall\, y \in (0,L); \qquad \forall\, t \in \mathbb{R}_+ \tag{A.14}$$

with

$$T(t) = [C_1 \cos(\omega\, t) + C_2 \sin(\omega\, t)]$$
$$Y(y) = [C_3 \cos(\lambda\, y) + C_4 \sin(\lambda\, y) + C_5 \cosh(\lambda\, y) + C_6 \sinh(\lambda\, y)]$$

whereby $T(t)$ is a time oscillation function with respect to the vibration modes, $Y(y)$ is a shape function of the vibration modes and C_1-C_6 are the integration constants. Considering the boundary conditions (A.15), the first three eigenvalues λ, natural angular frequencies ω and, finally, the natural frequencies for an Euler - Bernoulli beam f_{EB} can be determined (A.16). The numerical values for (A.16) are reported in Table A.1.

$$\mathrm{v}(0,t) = 0; \quad \frac{\partial \mathrm{v}(0,t)}{\partial y} = 0; \quad \frac{\partial^2 \mathrm{v}(L,t)}{\partial y^2} = 0; \quad \frac{\partial^3 \mathrm{v}(L,t)}{\partial y^3} = 0; \quad \forall\, t \in \mathbb{R}_+ \tag{A.15}$$

$$
\begin{aligned}
f_{EB_1} &= \frac{3.516}{2\,\pi} \cdot \sqrt{\frac{4\,E\,I_z}{\rho\,\pi\,d^2\,L^4}} \\
f_{EB_2} &= \frac{22.034}{2\,\pi} \cdot \sqrt{\frac{4\,E\,I_z}{\rho\,\pi\,d^2\,L^4}} \\
f_{EB_3} &= \frac{61.697}{2\,\pi} \cdot \sqrt{\frac{4\,E\,I_z}{\rho\,\pi\,d^2\,L^4}}
\end{aligned}
\tag{A.16}
$$

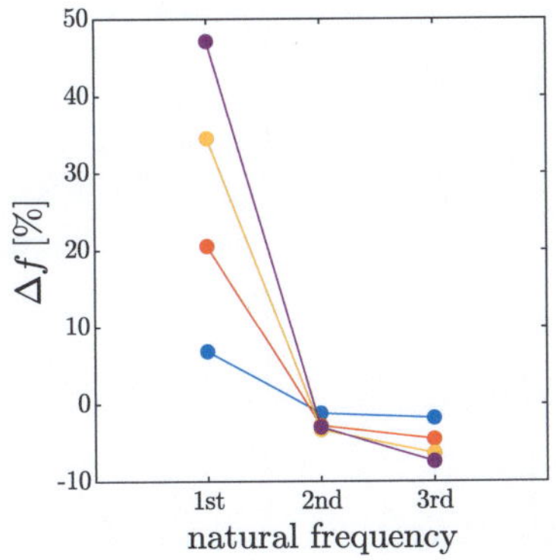

Figure A.7 – The relative change Δf of the first three natural frequencies is given by the difference of natural frequency of the free vibrating beam and the natural frequency of the loaded beam with a load of: blue- 0.025 N; red- 0.050 N; yellow- 0.075 N; magenta- 0.100 N. The absolute values are given in Table A.1.

The natural frequencies are calculated for a sensor shaft with the properties given in Table B.3. Figure A.7 and Table A.1 indicate that the natural frequencies are influenced by the level of pre-stress, here, represented by a load in x-direction. The natural frequencies are analyzed up to the third one because the values of higher natural frequencies strongly exceed the frequency range that is of interest for the present work. Since the actual orientation of the contact force changes in every time step and is not a priori known, this pre-analysis focuses on a constant load in x-direction for a qualitative evaluation of the dependence between deformation of the beam and corresponding natural frequencies. There is a strong change of the first natural frequency and weaker changes of the higher natural frequencies. The results of the finite element analysis are very close compared to the ones obtained by solving (A.13). Consequently, a discretization of the beam with 100 elements is adequate to include dynamical characteristics, e.g., natural frequencies but only under the condition that the time step in a transient analysis is sufficiently small. The influence of the stiffness proportional damping factor β, used in the finite element analysis, is neglectable because β is sufficiently small as indicated by the results shown in Table A.1.

Table A.1 – Overview of the first three natural frequencies of the beam models for various levels of pre-stress. The mode shapes are calculated using ANSYS.

Stationary shape	natural frequency 1st	2nd	3rd
	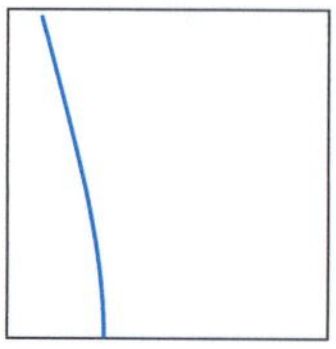	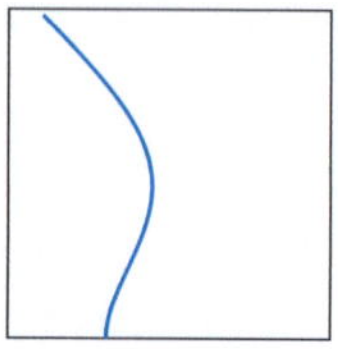	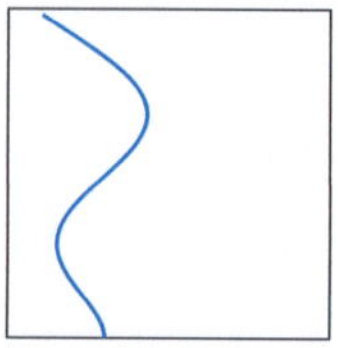
-	$f_{EB} = 28.67\,\text{Hz}$ $f_{FEM} = 28.66\,\text{Hz}$	$f_{EB} = 179.65\,\text{Hz}$ $f_{FEM} = 179.65\,\text{Hz}$	$f_{EB} = 503.02\,\text{Hz}$ $f_{FEM} = 503.20\,\text{Hz}$
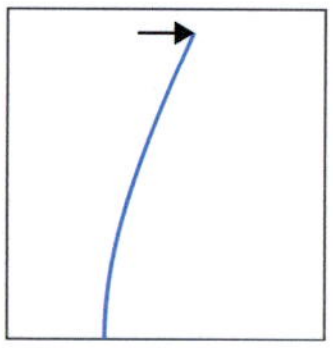	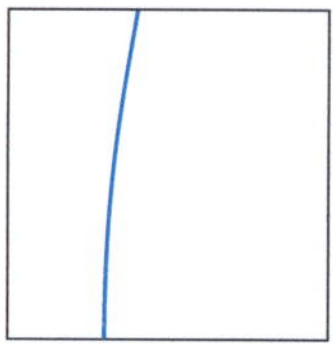	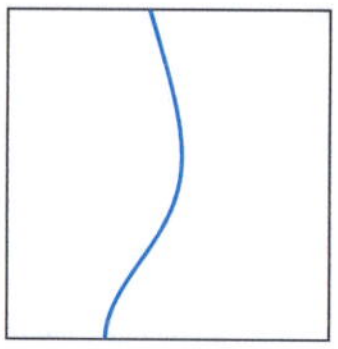	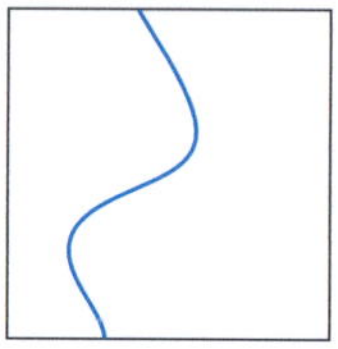
0.025 N	$f_{FEM} = 30.62\,\text{Hz}$	$f_{FEM} = 177.46\,\text{Hz}$	$f_{FEM} = 494.05\,\text{Hz}$
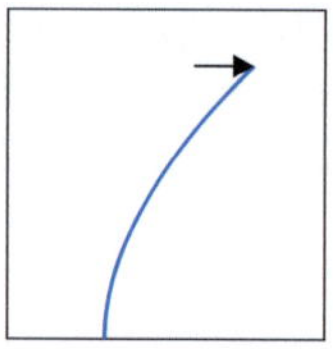	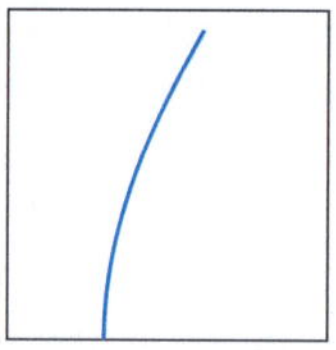	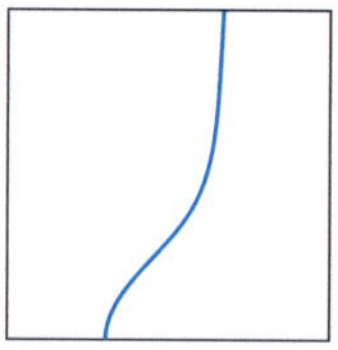	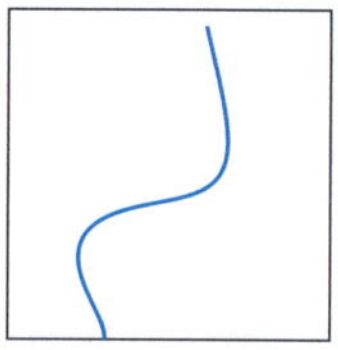
0.050 N	$f_{FEM} = 34.54\,\text{Hz}$	$f_{FEM} = 174.61\,\text{Hz}$	$f_{FEM} = 480.27\,\text{Hz}$
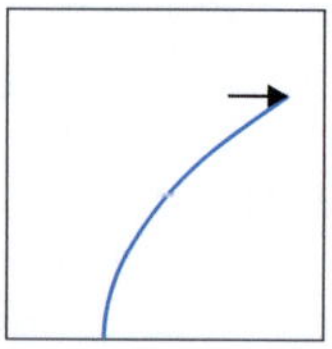	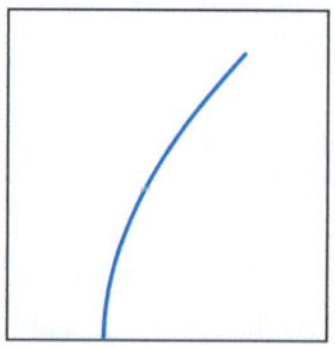	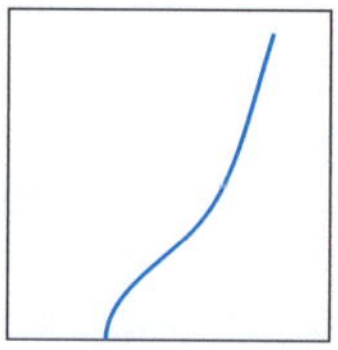	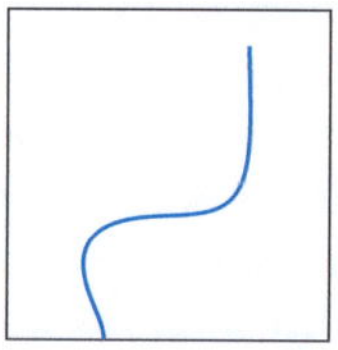
0.075 N	$f_{FEM} = 38.54\,\text{Hz}$	$f_{FEM} = 173.62\,\text{Hz}$	$f_{FEM} = 471.02\,\text{Hz}$
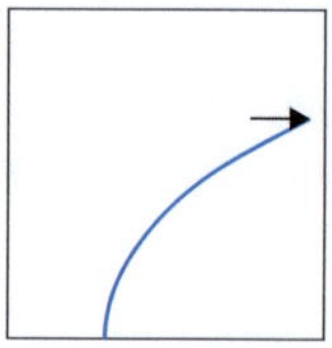	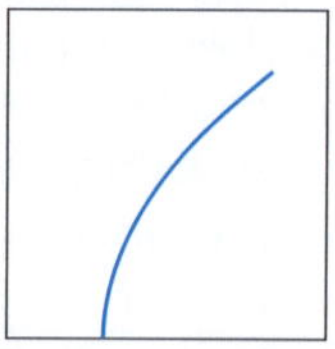	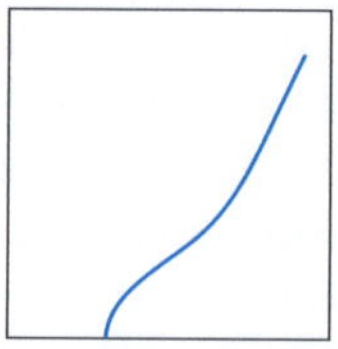	
0.100 N	$f_{FEM} = 42.17\,\text{Hz}$	$f_{FEM} = 174.22\,\text{Hz}$	$f_{FEM} = 465.76\,\text{Hz}$

B | Software & Hardware

This chapter provides all information regarding the used soft- and hardware, including the test objects, see Sections B.1, B.2 and B.2.2. The analysis procedure of the experimental data is described in Section B.3. Furthermore, the experimental setup is analyzed in terms of its dynamical characteristics, see Section B.4.

B.1 Software

The software mentioned in Table B.1 was used for theoretical and experimental analyzes.

Table B.1 – Overview of the hardware used in experiments.

Developer	Name	Version	Usage
Mathworks	MATLAB	2019b	numerical computation, data analysis, visualization
ANSYS	Mechanical APDL	2020 R1	numerical computation
Waterloo Maple	Maple	2018.2	algebraic computation
National Instruments	LabVIEW	2017	measurement data acquisition, motion control

The LabVIEW code used for measurement data acquisition and motion control of the connected axis (see Tabel B.2) is programmed by Florian Schale from Technsiche Universität Ilmenau. The author of the present work was only involved in the design process of the GUI and incorporated functionalities.

B.2 Experimental setup

The components and their arrangement are described in Subsection B.2.1. The specifications of the test objects and surfaces are given in Subsection B.2.2.

B.2.1 Hardware

All essential hardware components are given in Table B.2, and their arrangement is shown in Figure B.1. The experimental setup enables a repeatable positioning of the sensor and defined orientation of the test objects, and it also performs the scanning process in a suitable, repeatable way by an actuation of the x-guide.

In experiments, the sensor assembly is located above the test objects and the sensor shaft is orientated in the direction of gravity because of restrictions regarding the assembly of the included components. A guidance in x-direction displaces the sensor assembly during scanning. The distance to the test object is controlled by the y-guide. The z- guide is only used for automation of the experiment and is not involved in the scanning process. The sensor gets rotated by the φ_0-joint. The rotation of the

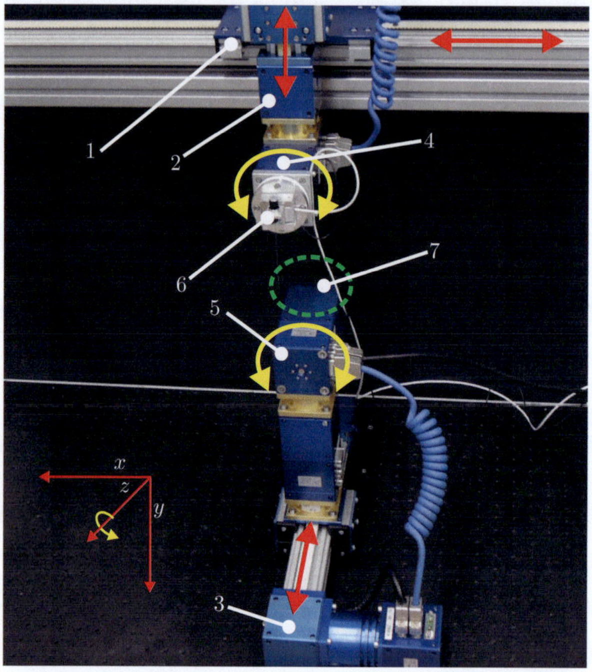

Figure B.1 – Experimental set-up: 1- x-guide; 2- y-guide; 3- z-guide; 4- φ_0-joint; 5- α-joint.; 6- sensor assembly; 7- object location. The sensor assembly 6 includes the torque sensor, force sensor, clamping, and sensor shaft. The test object is placed in the green highlighted area.

sensor around the z-axis. The test objects can be rotated around the z-axis using the α-joint. The α-joint is used to change the inclination of a plane contact surface, like in Section 5.3. The green-colored, dashed circle in Figure B.1 indicates the position of the test objects. The force sensor is serially connected to the torque sensor. So, if the torque sensor is loaded it deforms and rotated the force sensor around the z-axis. As mentioned in Chapter 4, this deformation is supposed to be neglectable as well as the deformation of the force sensor in consequence of a load because the influences on the recorded signals are small compared to other influences, e.g., the influences of the drive, see Section B.4. Therefore, the support of the sensor shaft is assumed to be rigid. The sensor shaft consists of a spring steel alloy, see Table B.3.

Table B.2 – Overview of the hardware used in experiments.

Function	Manufacturer	Type	Specification
torque sensor	ME-Meßsysteme	TD70	accuracy class 0.1, nominal load $\pm$ 0.05 Nm
force sensor	ME-Meßsysteme	K3D40	accuracy class 0.5, nominal load $\pm$ 2 N
amplifier	ME-Meßsysteme	GSV-1A4 M12/2	strain gauge amplifier, cut-off frequency 250 Hz
recorder	National Instruments	NI PXI 6221 M-Series	multifunction data acquisition device
x-guide	AMTEC	PLB090	positioning repeatability $\pm$ 0.05 mm, stroke 1 m
y-guide	AMTEC	PLB070	positioning repeatability $\pm$ 0.05 mm, stroke 0.2 m
z-guide	AMTEC	PLB070	positioning repeatability $\pm$ 0.05 mm, stroke 0.2 m
φ_0-joint	AMTEC	PR070	positioning repeatability $\pm$ 0.02°
α-joint	AMTEC	PR070	positioning repeatability $\pm$ 0.02°

Table B.3 – Properties of the sensor shaft.

Property	Value	Comment
d	0.4 mm $\pm$ 0.008 mm	[EN 10270-1]
L	0.1 m $\pm$0.001 m	actual dimension
material	spring steel	[EN 10270-1]
	$E = 206\,\mathrm{GPa}$	Appendix of [EN 10270-1]
	$G = 81.5\,\mathrm{GPa}$	Appendix of [EN 10270-1]
	$\nu \approx 0.26$	$\nu = \frac{E}{2 \cdot G} - 1$
	$\rho = 7850\,\frac{\mathrm{kg}}{\mathrm{m}^3}$	Appendix of [EN 10270-1]

It is clamped by a miniature jaw chuck which is mounted onto the force sensor. The actual length of the sensor shaft is determined by a conventional caliper. The parameters for the experiments are chosen to avoid that the sensor shaft undergoes plastic deformation and, simultaneously, to fit in the measuring range of the available equipment.

B.2.2 Test objects

All test objects regarding Sections 5.1 and 5.2 are manufactured using Ultimaker S5 3D-printer, see Table B.4. They consist of Ultimaker ABS Gray (Acrylnitril-Butadien-Styrol) which exhibits minimal deviations of the object shape as specified by the manufacturer. Therefore, it is supposed that the actual object is identical to the nominal one. The author of the present work is aware that the validity of this assumption is very limited, especially for 3D-printed objects.

> But, in the scope of the present work, the influences and consequences of small deviations regarding the object shapes are acceptable.

Table B.4 – Overview of the objects used in experiments. Note, the objects g_2 and h_2 are identical. The object shape functions are chosen based on pre-experiments (not documented in the present work).

Name	Shape function	Object
f_1	$f_1(x, \eta) = \eta + 8\,\mathrm{mm} \cdot \sin\left(\frac{\pi \cdot x}{50\,\mathrm{mm}}\right)$	
f_2	$f_2(x, \eta) = \eta + 8\,\mathrm{mm} \cdot \sin\left(\frac{\pi \cdot x}{50\,\mathrm{mm}}\right) + 1\,\mathrm{mm} \cdot \sin\left(\frac{\pi \cdot x}{2\,\mathrm{mm}}\right)$	
g_1	$g_1(x, \eta) = \eta + 0.5\,\mathrm{mm} \cdot \sin\left(\frac{\pi \cdot x}{2\,\mathrm{mm}}\right)$	
g_2	$g_2(x, \eta) = \eta + 1\,\mathrm{mm} \cdot \sin\left(\frac{\pi \cdot x}{2\,\mathrm{mm}}\right)$	
g_3	$g_3(x, \eta) = \eta + 1.5\,\mathrm{mm} \cdot \sin\left(\frac{\pi \cdot x}{2\,\mathrm{mm}}\right)$	
h_1	$h_1(x, \eta) = \eta + 1\,\mathrm{mm} \cdot \sin\left(\frac{\pi \cdot x}{1\,\mathrm{mm}}\right)$	
h_2	$h_2(x, \eta) = \eta + 1\,\mathrm{mm} \cdot \sin\left(\frac{\pi \cdot x}{2\,\mathrm{mm}}\right)$	
h_3	$h_3(x, \eta) = \eta + 1\,\mathrm{mm} \cdot \sin\left(\frac{\pi \cdot x}{3\,\mathrm{mm}}\right)$	
s	$s(x, \eta, \alpha) = \eta + \arctan(\alpha) \cdot x$	

For the experiments in Section 5.3 conventional sandpapers are used in order to represent different surfaces. The sandpapers are characterized by the average diameter of the grains as defined in [ISO 6344-2], [ISO 6344-3], see Table B.5. Additionally, images of the sandpaper surfaces are taken with a sixtyfold magnification. The images show that the sandpaper surfaces are roughly irregular with a stochastic distribution of the particles. The standards [ISO 6344-2] and [ISO 6344-3] define the limits for the geometrical dimensions of macro- and microscopic surface features in a different way than the present work.

Table B.5 – Overview of the sandpapers used in experiments in accordance to [ISO 6344-2], [ISO 6344-3]. The images are taken with a sixtyfold magnification.

Type	∅ particle diameter [μm]	Texture
P120	125	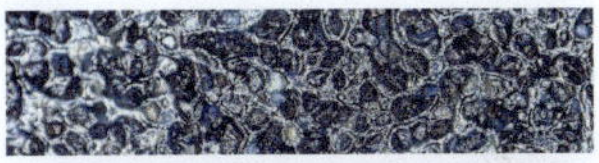
P180	82	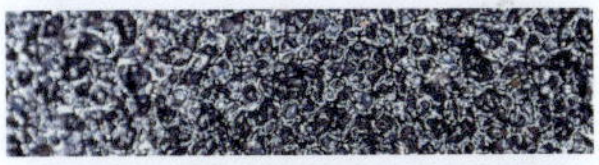
P320	46.2	
P600	25.8	
P1000	18.3	
P3000	6	

B.3 Evaluation of the experimental data

The data processing is applied for all experiments reported in Chapter 5. During an experiment, the time, support reactions, and support positions are recorded, see Section B.2. Each experiment is repeated five times. Thus, the following procedure is used to determine the mean value and corresponding standard deviation of the generated data. The recorded data is pre-processed before averaging, see Figure B.2.

- First, the recorded signals are centered by deducting the arithmetic mean of pre-signal sequence from the measured data. The pre-signal is recorded for 10 s while the sensor stays free in front of the test object.

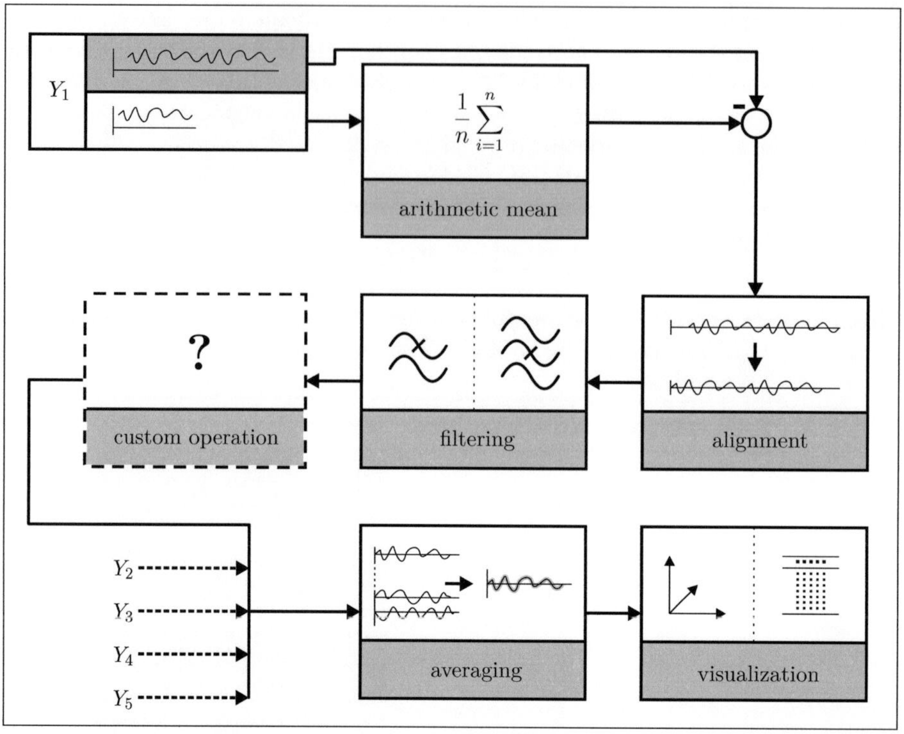

Figure B.2 – Data processing procedure. The experiments are repeated five times and the signals Y_1 to Y_5 result. The custom operation block belongs to specified procedures like the object shape recognition algorithm of [Steigenberger; 2013]. The custom operation block is explained in the regarding sections.

- Afterwards, the centered signal gets aligned using the position of the support as trigger. Furthermore, the signals are cropped with respect to the defined displacement distance of the support.
- The resulting signals are filtered in two steps by a low-pass and a band-stop filtering. The low-pass filters high-frequency components of, e.g., the motor. The band-stop filters the power line frequency (50 Hz hum). Both filters are of 6th order and characterized by a butterworth filter design (IIR-Filter). The filters are implemented in the MATLAB function `filtfilt`(·) to avoid a phase difference.
- In the next step, the data is treated by specified procedures, reported in Chapter 5. For example, in Section 5.3, the frequency spectra of the signals are determined applying a fast Fourier transformation (MATLAB function `fft`(·)).

- The resulting signals are combined by an ensemble average or simple arithmetic mean in order to determine the mean values and corresponding standard deviations in every time step.

- Finally, the data gets depicted in order to show the results and support the discussion.

Additionally, selected data was evaluated using the guide to the expression of uncertainty in measurement [GUM:1995]. For example, the uncertainty of reconstructed object shapes was exemplarily analyzed using a Monte-Carlo simulation. But, because of the fundamental character of the present investigation, this expensive method is not used. Nevertheless, the results of the Monte-Carlo simulation demonstrated that the uncertainties of the determined values are in a reasonable magnitude.

B.4 Dynamical characteristics of the experimental setup

The interaction between sensor shaft and test object/surface is analyzed in Chapter 5. Two sets of pre-experiments are done to avoid misleading interpretation of the captured data, see Tables B.6 and B.7. The goal of these pre-experiments is on one hand to demonstrate the influences of the data processing explained in Figure B.2, and on the other to evaluate the response of the sensor in consequence of a displacement. Based on this evaluation an adequate displacement velocity must be chosen.

B.4.1 Frequency response of the structural components

The first set of pre-experiments are summarized in Table B.6. The sensor assembly is gradually completed to determine the influences of the different structural components on the captured signals, compare Figure B.1. First, in configuration A, only the torque and force sensor are connected to the guides. Afterward, in configuration B, the clamping is added and, finally, also the sensor shaft (configuration C). Each configuration is tested for a translational displacement of the x-guide with $v \in \{0\,\mathrm{m\,s^{-1}};\ 0.001\,\mathrm{m\,s^{-1}};\ 0.005\,\mathrm{m\,s^{-1}};\ 0.01\,\mathrm{m\,s^{-1}}\}$. The signals are recorded with a sampling rate of 500 Hz and for configuration A as well with 1000 Hz. The data belonging to 1000 Hz are analyzed but not included in the present work since a sampling rate of 500 Hz is suitable for the present experimental setup. The natural frequencies of the sensor components are analyzed by exciting the system with a jerk caused by a rotation of the φ_0-joint.

Table B.6 – Overview of the first set pre-experiments to characterize the dynamical properties of the experimental setup. The configuration refers to the connected components: A- torque and force sensor only; B- torque and force sensor and clamping; C- torque and force sensor, clamping, and sensor shaft. The columns Start and End indicates the start and end position of the sensor with respect to the x-axis in case of a translational velocity v or in the case of an angular velocity $\dot{\varphi}_0$ to the φ_0-joint. The experiments are repeated three times, the "X„ indicates the number of the run.

Nr.	Configuration	$v\left[\frac{m}{s}\right]$	Start [m]	End [m]	SR [Hz]	Name
1	A	0.000	0.500	0.500	500	VV_A_0_500_00X
2	A	0.001	0.425	0.575	500	VV_A_0_500_00X
3	A	0.005	0.425	0.575	500	VV_A_0_500_00X
4	A	0.010	0.425	0.575	500	VV_A_0_500_00X
5	B	0.000	0.500	0.500	500	VV_B_0_500_00X
6	B	0.001	0.425	0.575	500	VV_B_0_500_00X
7	B	0.005	0.425	0.575	500	VV_B_0_500_00X
8	B	0.010	0.425	0.575	500	VV_B_0_500_00X
9	C	0.000	0.500	0.500	500	VV_C_0_500_00X
10	C	0.001	0.425	0.575	500	VV_C_0_500_00X
11	C	0.005	0.425	0.575	500	VV_C_0_500_00X
12	C	0.010	0.425	0.575	500	VV_C_0_500_00X
13	A	0.000	0.500	0.500	1000	VV_A_0_1000_00X
14	A	0.001	0.425	0.575	1000	VV_A_0_1000_00X
15	A	0.005	0.425	0.575	1000	VV_A_0_1000_00X
16	A	0.010	0.425	0.575	1000	VV_A_0_1000_00X
		$\dot{\varphi}_0$ [°/s]	Start [°]	End [°]		
17	A	30	1	0	1000	VV_A_EKF_00X
18	B	30	1	0	1000	VV_B_EKF_00X
19	C	30	1	0	1000	VV_C_EKF_00X

Every pre-experiment is repeated three times to guarantee minimal statistical testing. The time-series signals of all configurations for a velocity of $0.001\,\mathrm{m\,s^{-1}}$ are shown in Figure B.3. The blue signals are only centered and aligned. They are not filtered excepting the influences of the amplifier and the corresponding cut-off frequency, see Table B.2. The red signals are filtered as described in Section B.3 using 100 Hz as cut-off frequency for the low-pass filter.

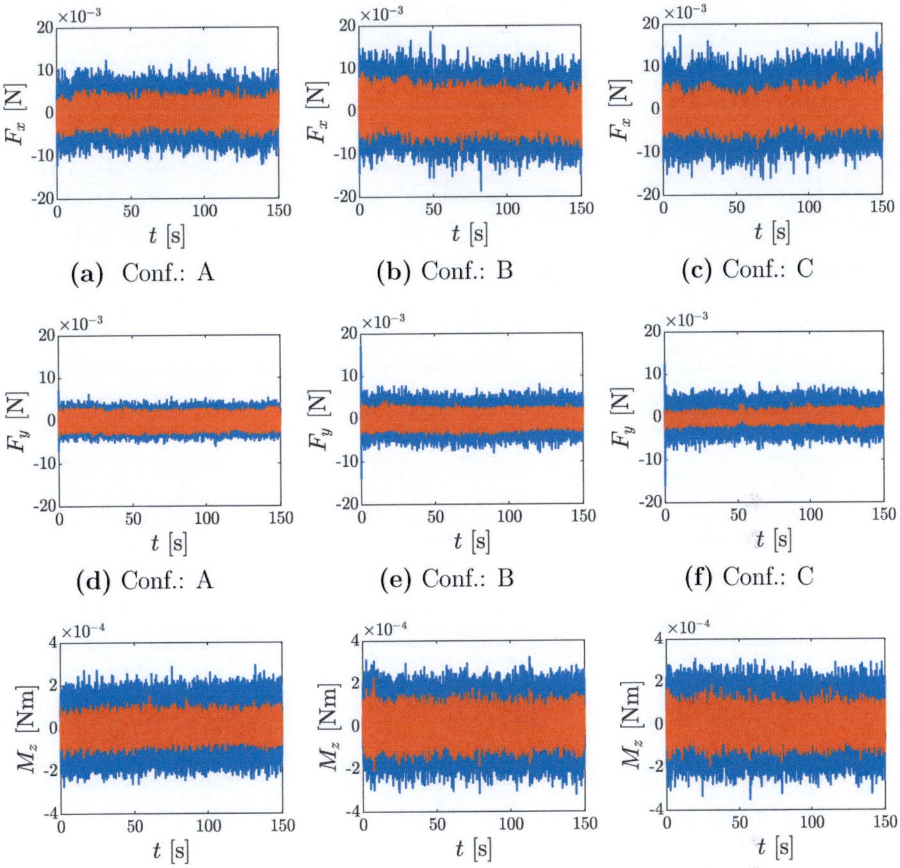

(a) Conf.: A (b) Conf.: B (c) Conf.: C

(d) Conf.: A (e) Conf.: B (f) Conf.: C

(g) Conf.: A (h) Conf.: B (i) Conf.: C

Figure B.3 – Time series signals for F_x, F_y, M_z for $v = 0.001\,\mathrm{m\,s^{-1}}$. In all subfigures, the abscissa indicates the time in s and the ordinate the amplitude of the corresponding signal: blue- raw signal; red- filtered signal.

As expected, the amplitudes of the filtered signals get reduced compared to the ones of the raw signals. Comparing the three configurations, the amplitude of the raw signals slightly increase from configuration A to C, but the amplitudes of the filtered signals remain constant, except the signals belonging to M_z where nearly no changes are observable in both cases. The signals for F_x show in all cases a larger amplitude as the ones for F_y. Figure B.4 illustrates the frequency spectra of F_x, F_y, and M_z, in a top-view perspective, for all repetitions, configurations and velocities, compare Table B.6.

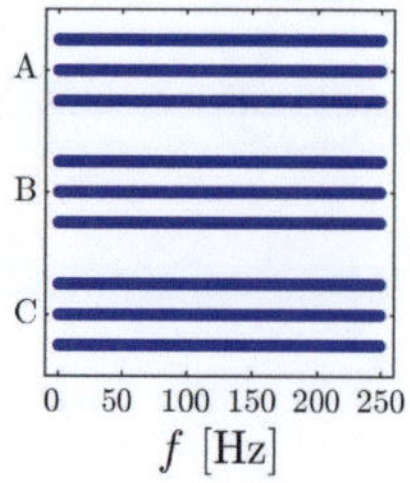

(a) F_x; $v = 0.0\,\mathrm{m\,s^{-1}}$.

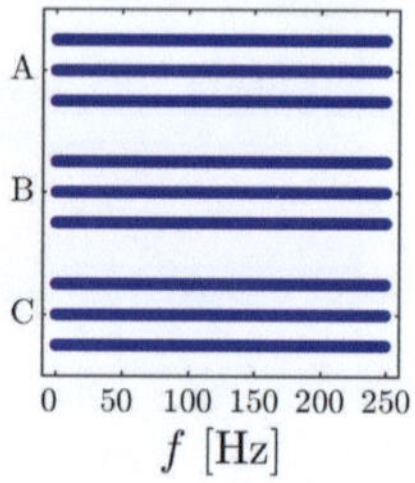

(b) F_y; $v = 0.0\,\mathrm{m\,s^{-1}}$.

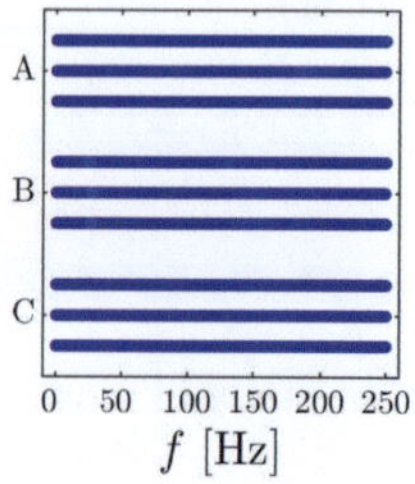

(c) M_z; $v = 0.0\,\mathrm{m\,s^{-1}}$.

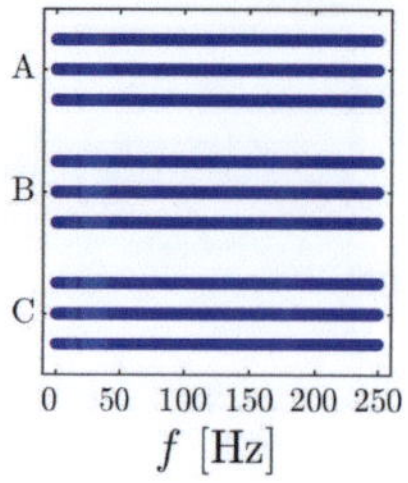

(d) F_x; $v = 1.0\,\mathrm{m\,s^{-1}}$.

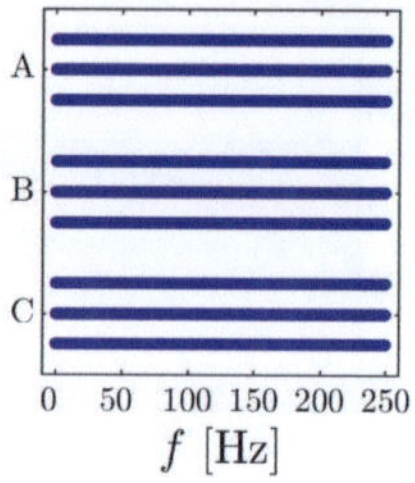

(e) F_y; $v = 1.0\,\mathrm{m\,s^{-1}}$.

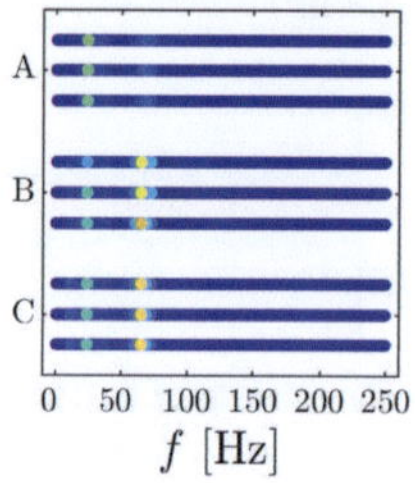

(f) M_z; $v = 1.0\,\mathrm{m\,s^{-1}}$.

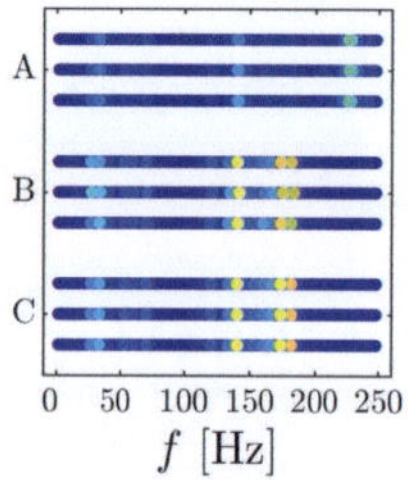

(g) F_x; $v = 5.0\,\mathrm{m\,s^{-1}}$.

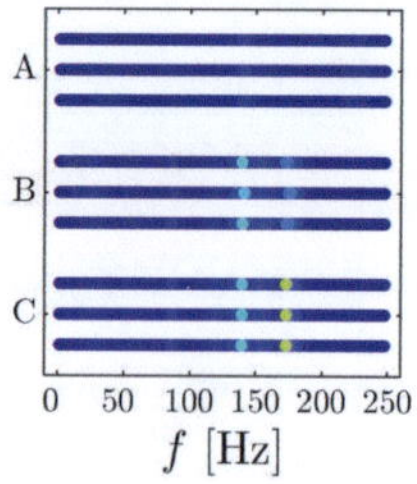

(h) F_y; $v = 5.0\,\mathrm{m\,s^{-1}}$.

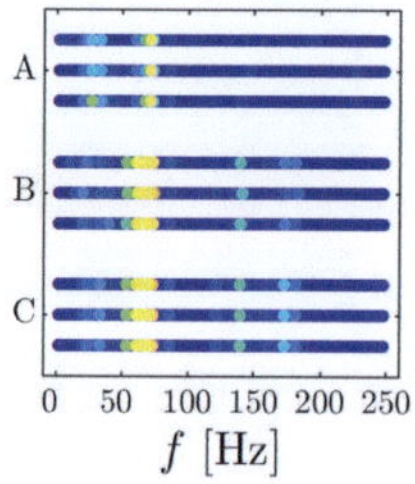

(i) M_z; $v = 5.0\,\mathrm{m\,s^{-1}}$.

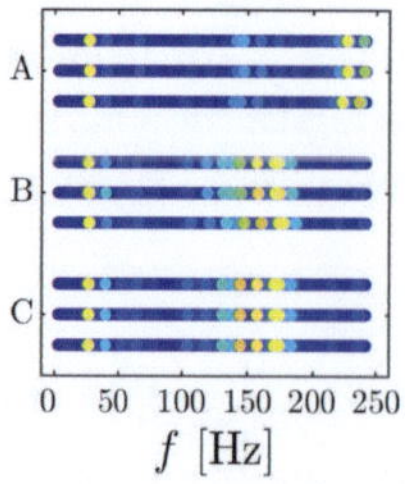

(j) F_x; $v = 10\,\mathrm{m\,s^{-1}}$.

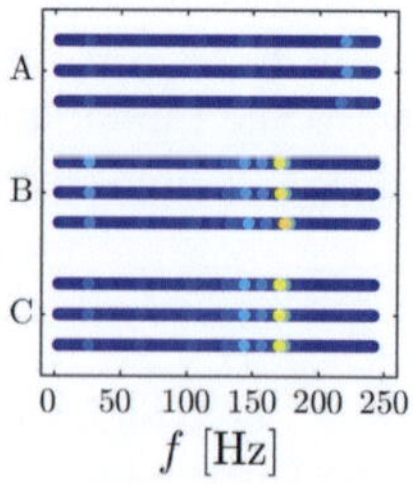

(k) F_y; $v = 10\,\mathrm{m\,s^{-1}}$.

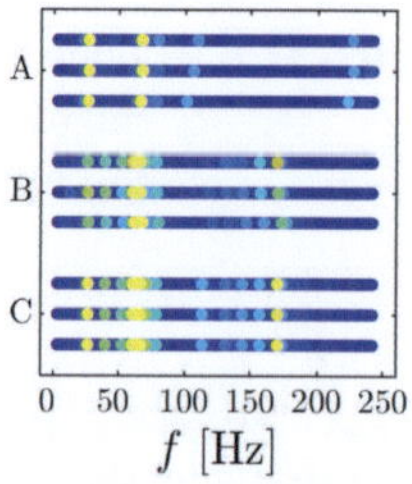

(l) M_z; $v = 10\,\mathrm{m\,s^{-1}}$.

Figure B.4 – Frequency spectra for F_x, F_y and M_z belonging to the experiments 1-12, see Table B.6.

The plots must be interpreted as follows: the abscissa indicates the frequency, the ordinate shows the configuration and the corresponding three runs, so each *line* is one repetition, the colors correspond to the amplitude of the vibration, whereby dark blue indicates minimal and yellow maximal amplitudes. The signals are processed as described in Section B.3 but no low-pass filter is applied. It holds true for all signals that for $v = 0\,\mathrm{m\,s^{-1}}$ there is no dominant frequency and that the runs within in one parameter set are similar. For F_x and F_y there are notable frequencies of approximately ≈30 Hz, ≈140 Hz, and ≈175 Hz. The amplitude of these frequencies increases with increasing v. In the cases of F_x and F_y for $v = 5\,\mathrm{m\,s^{-1}}$ and $v = 10\,\mathrm{m\,s^{-1}}$, there are changes in the spectra comparing configuration A and B, see Figure B.4 (g), (j) and (h), (k). Here, for configuration A there is a frequency at ≈240 Hz, and for configuration B at ≈175 Hz. This change belongs to the presence of the clamping. In the case of M_z, there are three major frequencies: ≈30 Hz, ≈70 Hz, and ≈175 Hz. The largest amplitude corresponds always to a frequency of ≈70 Hz, and the frequency of ≈175 Hz appears in consequence of the added clamping. Generally, some of the mentioned frequencies can be found in most spectra, but the amplitudes differ because of the sensitivity of the force and torque sensor to the current direction of excitation of the sensor. For configuration C, there are no notable changes compared to configuration B in all cases.
Comparing the spectra in Figure B.5, it becomes notable that the frequencies ≈140 Hz, and ≈175 Hz in the case of F_x and F_y and ≈10 Hz, as well as ≈70 Hz for M_z, correspond to influences of any component of the experimental setup, e.g., ≈175 Hz to the clamping. Whereby the frequency of ≈30 Hz which is observed in most signals in Figure B.4 seems to be related to the actuation/motor of the x-guide because it is not observable in Figure B.5.
In no experiment, the natural frequencies of the sensor shaft can be clearly distinguished. Moreover, the first natural frequency of the sensor shaft seems to be close to the frequency that presumably belongs to the motor, see Subsection A.3. In a final series of pre-experiments, decay tests are used to determine the natural frequency of the sensor shaft.

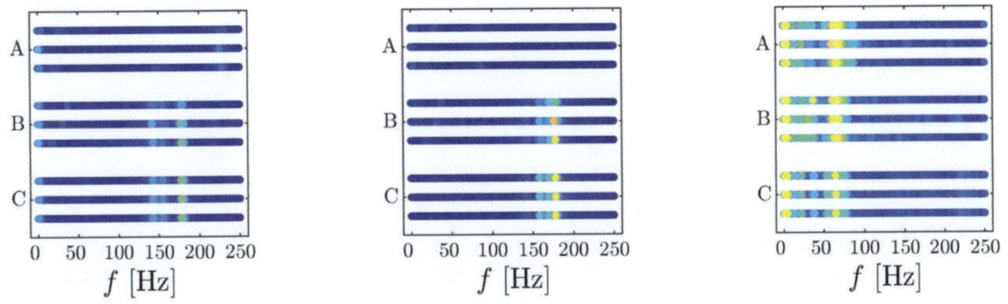

(a) F_x (b) F_y (c) M_z

Figure B.5 – Frequency spectra for F_x, F_y and M_z belonging to the experiments 17-19, see Table B.6.

B.4.2 Decay test

For the decay test all components are mounted (configuration C). The sensor is translationally displaced with a velocity of $v = 0.001\,\mathrm{m\,s^{-1}}$ while the sensor shaft gets pulled along an edge that is placed in a distance η from the clamping, see Table B.7.

Table B.7 – Overview of the decay tests. The decay tests are repeated five times, the "X„ indicates the number of the run.

Nr.	Configuration	$v\left[\frac{\mathrm{m}}{\mathrm{s}}\right]$	η [m]	Sampling rate [Hz]	Name
1	C	0.001	0.080	500	E_0065_0_0_0001_00X
2	C	0.001	0.080	500	E_008_0_0_0001_00X

When the sensor shaft loses contact from the edge, a huge snap-off takes place resulting in an optical notable oscillation of the sensor shaft. This vibration is also indicated by the recorded support reactions, see Figure B.6. In all cases, the amplitude of the vibration is decaying. The initial amplitudes are larger for $\eta = 0.065\,\mathrm{m}$ than for $\eta = 0.08\,\mathrm{m}$, because of the closer distance of the edge and corresponding larger deformation of the sensor shaft. The spectra of all signals are illustrated in Figure B.7.

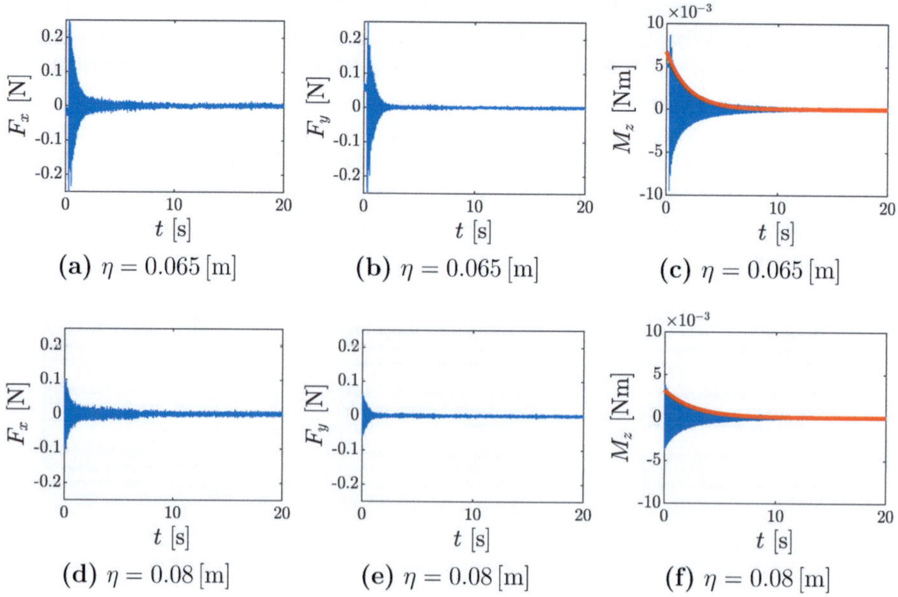

Figure B.6 – The time series signals of one run for different η are colored in blue. The red line corresponds to the upper enveloping curve.

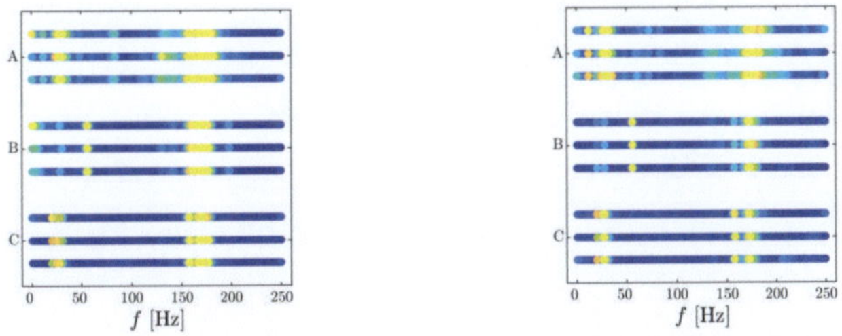

(a) $\eta = 0.065\,[\text{m}]$ **(b)** $\eta = 0.08\,[\text{m}]$

Figure B.7 – The spectra for all decay tests described in Table B.7.

In each spectrum, frequencies of $\approx$ 27 Hz-30 Hz and $\approx$ 160 Hz-173 Hz can be observed. Furthermore, in the case of F_y there is a frequency at $\approx$ 55 Hz. The frequencies at $\approx$ 27 Hz-30 Hz and $\approx$ 160 Hz-173 Hz are close to the natural frequencies calculated for the sensor shaft in Section A.3 and to the ones discussed in Subsection B.4.1. But due to the characteristics of the time series signals and the clearer peaks in the spectra compared to the spectra in Figures B.4 and B.5, it is reasonable that in the present case the natural frequencies of the wire dominate the spectra. The first natural frequency in the spectra is determined to be $\approx$ 27.70 Hz and the second one to be $\approx$ 171.00 Hz. Both values are close to the theoretical results in Section A.3 with 28.67 Hz and 179.65 Hz. Finally, the damping ratio is determined with respect to the first natural frequency and the signal of M_z, see Figure B.6. A value of 0.0027 is determined as a representation of material as well as structural damping.